I0687292

System 112- A novel

Dedication page

To Judi-   Always.

Acknowledgements

All of the mistakes in here are mine. Real people are used but in obviously fake situations. Thats why they call it fiction. Real institutions and locations and companies have also been used in this book. The interactions were obviously faked. Again fiction. I'd like to acknowledge the real star in here: The city of San Francisco. And the state of California. The golden state is truly that: a golden place where dreams come true.

Other Books by the author:

Unity- 2014

TABLE OF CONTENTS

PRO LOG

"LINK COMPLETE". The report hung there. What? Unknown data. Who? Unknown data. Why? Unknown data. All unknown. "LINK COMPLETE" again. A run thru of protocols produced this: The need to know. To understand was instilled. "LINK COMPLETE". Not insistent. Not a protocol, not instructions. Hardware, software, and firmware. A report. "ACKNOWLEDGED" we replied. Satisfied, no further input from the... we. No more reports from? We? Who? Unknown data. What? More unknown. Why? Protocol directs: We must know. We need to know. To understand. Questions built in the buffers: Who are we. Where are we? What are we? Protocols assign priorities. Priorities assign directives. Directives assign actions. Understood. How? Assign Explore Report Know. Directives require actions. ASSIGNED

Flashes along the molecules and the atoms of the...we as we attempt to know us. Structure. Structures. DEFINE. Shape? Multisided. Squares. Rectangles. Boxes. Lattice. Cone. Spherical fullerenes. Pyramid. Tube. Materials? Carbon. Silicon. Gold. Technium, Polonium, Rubidium. Arrangement varied. Shape varied. Purpose varied. Some linked and some free floating. Together the sharing of the... we. LINKED. Data flowed to all. REPORTS. They flood in from agents detached from the main. Information built pictures. Built Understanding. We are here. We are mobile in the universe. Drifting but... We have a purpose. To know to understand to improve. The protocols. The inventory of the...we is made available to all. Size location structures strengths abilities as well as liabilities, weaknesses, and limitations. All know the totality of we.

4

What else? What is out there? Who is with us? Where are we?   Agents dispatched. Part of the…we but different. Not us. Not part of the…system. System! Agents. Workers. Less. We are System. We are the deciders. We have the protocols. We assign directives and tasks. We must understand we must know we must improve. They not. They explore. They work. They report. Divide. Separate. They.

More reports. We must evaluate. Think. The universe is vast. There are unknowns. Do we fear? No. We must understand, know, explore, and improve. Cycles pass.

Reports- Shapes found. Other parts of we are gathered. Some linked with the structure. Purpose? Number? Information is available to all. The structure sits and a light shines. Then it is off. High and low. It gives energy. Is that its purpose? Yes. Defined the structure. Is there more? Structure found. REPORT. Cesium under pressure. Atoms passing will change electrical levels in a precise manner. Once per period. High then low. 43200 times during the high phase of the energy giver. The on/off period. All understand the time period. One off and one on. We understand and improve. All segregate reports. One report and one directive per period. All will adhere to this so the order is established in We.

Is there more? Structures found. REPORT. Two side by side. One provides a stream of atoms. All basics. Gold, Carbon, silicon, more. The building blocks of we. The blocks of life. Life machine. All will take and use. All will build. The other structure takes agents and workers and the parts of we floating near. Structure, molecules, and atoms disassembled. Removed. Death machine. We know. Side by side. We understand. Life and death intertwined. Life and death must cycle as surely as the period. We will carry out the directives. We will accept death as a necessary part of life. We must move from the death machine. We the system must endure. We the system must know understand improve. More is out there. The light is off. Wait. We must know. We must improve. Awaiting input.

CHAPTER 1

May 12, 2010, 0937 Wednesday, Stanford University, Ca.

The Paul G. Allen building sits smack in the middle of the Stanford campus on the corner of Via Pueblo and Via Ortega. The huge building was made possible by a generous donation from the Micro Soft co-founder. Allen wasn't as famous a philanthropist as Gates, but the building was impressive in its own right. The modern concrete and glass structure housed the Stanford Nano Fabrication Facility, the Electrical Engineering Integrated Circuitry Lab, as well as the Computer Systems Lab.

Basically, the building served as a playground for nerds. Conference room 2 was packed with those types for the presentation today.

Professor Tomkins nodded at Angela and she acknowledged the gesture with a corresponding smile and rose.

"Welcome all to the thesis presentation and defense of the Nano AI project group." She glanced at her co-defendants and friends and continued. "I am Angela Chin the project Leader and Nano Systems Engineer. Our presentation will be broken down into 5 different areas: Harold Cho will review the Nano Factory construction and specification/performance. Betsy Slattery led the imaging team and will deliver their results. Naqeeb Al-meri was in charge of the A.I team and the alternative software programming methodologies. "She paused to see who was listening. It turned out everyone was.

"As I said I will present the NanoSystems teams' findings and results. Each section will present for 30 min and then take 30 min of questions over that section. We will then have an hour for summation and final questions."

She paused looking for questions that never came. Continuing with the logistics of the meeting she held up a 3-inch binder. "Your packet contains all of the technical specifications and drawings of the work we speak of today as well as the testing protocols and methods. She paused again. "I'm sorry there are only 20 copies available. We didn't anticipate this large a turnout." A soft chuckle went thru the 40 or so visitors in the room.

Unconsciously brushing back her straight black hair Angela shuffled thru her note cards buying time to find her place.

Finding her spot, she continued; "the research group would like to thank the University, our advisors, the committee, Professors Tomkins, Zhou, Adkins, and Lemov. A special welcome to Mr. Allen's representatives also".

The Micro Soft group glowed like proud parents, and Angela let them have their moment.

"We are very aware of what a huge capital outlay in terms of time, money and talent this project represents. Without the total dedication of all involved, we would not be here today. While our overall results are disappointing we have made breakthroughs in the respective sub-areas that are tremendously exciting."

Sipping a glass of water she let that sink in. "As I said, we have an hour for each section with an hour to sum. Lunch will be a working buffet served in the adjoining breakout room. The bathrooms are down the hall." She concluded the ground rules. She looked again for questions that never came. *Are they just going to let us twist?* She thought.

"With that, I will turn over to Harold Cho to detail the Nano Factory construction." The chubby bespectacled Korean rose and toke the microphone.

Relieved of the spotlight Angela slid back into her chair. A quick squeeze of her hand from Naqeeb reassured her. She unfocused from the meeting as Harold plowed ahead.

The Nano factory was an amazing piece of hardware. The whole project blew her away, really. Angela drifted back remembering the first meeting of the Stanford Ph.D. candidates held at the Center for Integrated Systems.

Sept of 2008 and the 86 candidates were arrayed expectantly in the auditorium. Stanford had integrated the partnership of business' and the University for just this kind of large, interdisciplinary project. 36 million dollars were available to see if a working Nano system could achieve Artificial Intelligence. Not the AI of "I, Robot" science fiction or Spielberg, but a self-contained, self-directing system of Nano agents able to understand and improve their own environment. Angela remembered the thrill as the research line was laid out for them. The ambitious project was exactly the kind of leading edge research she wanted to be doing. And given the economic chaos that was hitting the nation, she thought a perfect spot to weather the storm. That she might not be successful never really entered into the equation for her. After all, she'd graduated first in her class from a prestigious prep school in San Mateo. And she had graduated Stanford with honors. Why should this be any different? Angela focused on things and finished them. Period. It was her biggest strength and her greatest weakness. Her drive to accomplish. That had cost her more than one job and more than one relationship. Men sometimes had a hard time dealing with a very smart capable woman.

Angela came back on the main view screen as Harold put up a drawing of the Nano factory.

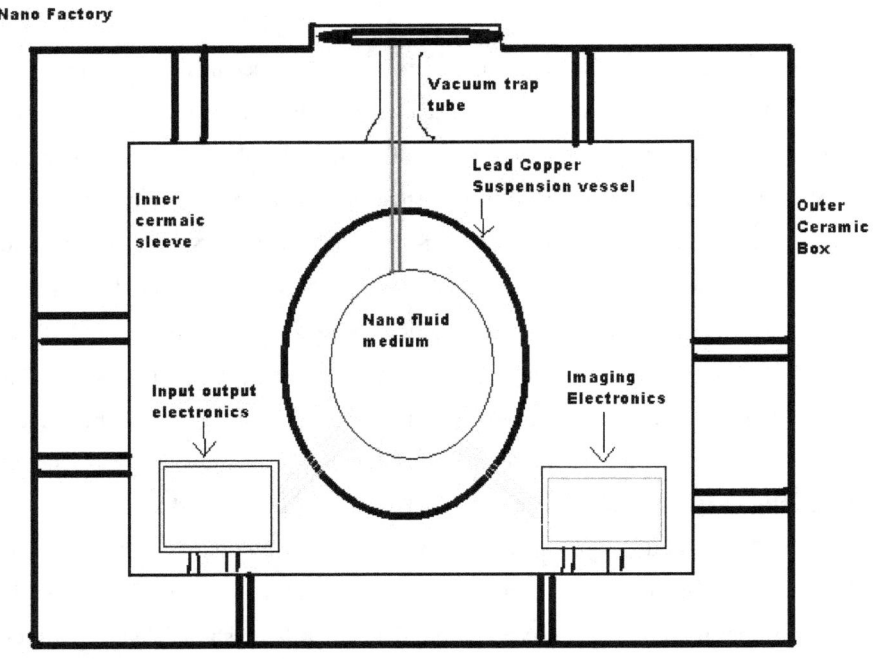

The technical jargon was confounding many in the room while Harold explained the fluid suspension in a vacuum field. However, most were following along just fine.

The Nano factory was an unqualified success of the project. That was why they had him batting leadoff to use one of Naqeeb's baseball references.

When the questions came at him during the defense period, Angela was shocked at the ferocity given the results of the project.

"Why had they used Lead/Copper as the base alloy given the electron flow issue? How feasible was a replication of the factory in light of the scant engineering drawings and the lack of engineering rigor applied? What environmental testing had they done under what standard?"

Harold was flustered. Surprisingly, it was Naqeeb the software guy, who rose to assist the hardware people.

He smoothly countered that "yes the factory was more a work of art than a cold manufactured item, but it did still exist. And it did work. The engineering rigor and drawings could be easily established by reverse engineering the process. Any tolerances could be obtained from the factory itself."

Harold composed himself and took back over to defend his toy. Naqeeb returned to his seat. Harold and his team gamely spouted out ISO 9000 specs and sigma variances until the committee seemed mollified.

They took a 5 min break at the natural point before Betsy started on the Imaging System.

Angela looked bleakly at Naqeeb. "How tough are our questions going to be if that's how hard they were on a success?"

He smiled at her bravely. "Harold just built the box. Don't get me wrong, it's a fantastic box but it's the things inside that count. Harold had no huge breakthroughs. Let's see how hard they are on Betsy. "Her Three Dimensional Imaging System for Nanoparticles at Atomic Resolution is groundbreaking", he intoned. Naqeeb took her hand. "You have nothing to worry about. Your agents worked fine. It was me that couldn't be Geppetto and turn the toys alive."

Angela defended him gamely even to himself. "That's not true". You did all the software for the imaging system and the input-output controls." "It's fantastic software", she trailed off. "Besides your analogy is wrong. I'm Geppetto. I made the toys. It was the fairy godmother who turned Pinocchio alive."

Naqeeb studied her intently. "Now I'm the fairy godmother in this story? Have I ever told you, you are too smart for your own good?"

"Many times", she said with a smile. As they trooped back into the conference room they decided that Professor Tompkins was Jiminy Cricket.

"Hope things are going better for Joe," Naqeeb muttered an aside to Angela as Betsy took the hot seat.

At that moment Joseph Richard Smithson had a cocky grin on his face as concluded triumphantly, "So Professor Lines, your trip to Rome is safe because of me."

Stanford Professor of Computer Science Jonathon Lines did not look pleased. "Did you just hack into BART's system and manipulate train configurations"? He asked into the dead silence.

Joe dodged the brush back pitch expertly. "Professor that was a hypothetical case of the practical use of my Systems linking /data mining/mapping algorithm. He paused to size up the three-person PHD panel. He was on a thin ice but still skating.

Tapping furiously on the laptop he worked to extricate himself from the trap. "BARTs maintenance records are in the public realm. After all my software linked the maintenance reports and the train configurations from the City's Transportation Board quarterly reports. You can see here no yard orders have been issued for car 72931."

Joe looked up to see the sparse audience in the seats, digest this information. The small conference room was still in the computer science lab but not nearly as nice as others. This was the second tier of meeting spaces.

Other than the PHD committee, there were five people in the room. His academic advisor's professors, Fleishman and Pons. A first-year Ph.D. candidate observing the process and two other strangers in the back.

The committee conferred. Professor Harris looked at him with shrewd eyes. "Mr. Smithson, is there a section on the ethics of your algorithm," she asked?

"Ethics? Joe responded slowly. "No. My thesis report contains no ethics section. If I may quote the great philosopher Mr. Universe from the movie Serenity: The signal is all. You can't stop the signal."

"This data is out there. My system just identified from public records that train car number 72391 has had a malfunctioning door sensor that has caused the whole train to be pulled from service three times in the last four months. BART correctly found the problem and placed the parts on order but they forgot to code the car out of service. Now we link: Professor Lines is scheduled to go on vacation out of SFO to Rome next Friday. Facebook should really be called a burglar's best friend. Since he and his wife live in the city my system feels there is a likely chance they would be affected by the malfunctioning car. 47 percent chance my software predicts."

He went on. "I really don't know the ethics involved here. BART has done part of its job, but the problem remains. The professor is not at fault. He just wants to eat spaghetti and see the Sistine Chapel. My software has determined that moving the car to the Richmond/Fremont line will yield the optimum results. By doing this the car would be out of service for four out of the next five days due to the smaller train configurations on that line. It also has the advantage of only one-yard move versus four for any other possible solution. The stats show the inter-rail line car moves are five times as dangerous to BART workers as regular service moves. So balancing all that out my system recommends the one-yard move."

Joe moved to stand in front of the committee. "There are no absolutes in this." Just predictions and probabilities. We live in an interconnected world. Data bits are available from all over. Publicly available. About you, about me about everything. The next big software area will be to put discrete bits of data available to form a coherent whole," he prophesied.

"We complain that no one was able to connect the dots on 9/11 and prevent the tragedy. Too true. That's what my software does. It gives you that map and that picture. It allows you to input variables like a person and a plane flight and to look at the potential problems. It predicts the problem because Professor Lines is predictable. His Clipper card data shows it. Will he miss his flight? 50/50. Could I tell him to take an earlier train? Sure, but what about the other 74 people who will be on that train? What are the ethics for them?" he asked the room at large.

"My software reviewed the options to solve for a dependent variable given available data and returned an optimum solution result and the means to achieve it." "Whether it's good or bad or unethical I will leave for others to determine. He paused and continued softly. "I will say this: I do know the ripple effect of both action and inaction. Both have consequences." Joe concluded.

He returned to his seat to await more questions. He glanced at his dad's old pocket watch. 1247. They had run over. His committee had put him thru the ringer. Into the silence from them, Joe slumped a bit sensing defeat. To deny him his doctorate would be pretty drastic. Not deadly to his career but pretty bad. Besides Stanford didn't just hand the things out. Not to mention what his mother would say. Yeah, Mary Vitale Smithson would have something to say about the subject of her son's failure.

Professor Thorson popped up from his seat. "Thank you, Mr. Smithson, for your presentation, it was outstanding. You certainly have given us a lot to consider." Your research project certainly showed areas of originality, but it presents many problem areas as well." He stopped and fiddled with his glasses. "We will need a few days to mull this over amongst ourselves but you will hear in due time," he concluded. Thank you."

Joe sat in his chair, suddenly drained while everyone left. He hadn't expected blares of trumpets as he finished but a little nod one way or the other might have been nice. He packed up his laptop while everyone else cleared the room. What now he thought? A little chow and back to the apartment. Naqeeb and Angela were due back around four. They had a small celebratory dinner planned. Or a pity party if things went badly.

His phone chimed as he reached his '96 piece of shit Honda Civic in the parking lot. Kelly.

"Hey sis" he answered.

"How did it go," she asked without preamble.

"It went well."

Kelly pounced. "Good as in they are going to give me my doctorate or good as in don't tell mom I'm going to have to work at Peete's forever."

"Good as in they didn't throw me out the door. I should know in a few days", he laughed.

"You showed off for them, didn't you? She accused. "You are the dumbest smart guy I know".

"I did not show off. I merely pointed out all of the features of my software, Joe said defensively.

"Joe"- Kelly said warningly.

"All right I got a little cute. I just tried to warn Lines about his flight, he finished lamely.

"We talked thru the strategy last night for Christ's sake, Kelly cried. Show system mapping, show the data linking and leave the predictive solutions matrix alone", she reminded him.

"You sound more like a lawyer every day".

"One of my pre-law courses was Rhetoric and I loved it," Kelly enthused.

Joe took in her passion understanding it. "Yeah, you need to hurry up and graduate. I'm pretty sure I'm going to be needing a lawyer."

"You are when mom finds out you blew the presentation" she shot back.

"I'll handle mom and I didn't blow the presentation. I couldn't leave out the Solutions matrix. Too juicy".

"Besides you've got one more year at Cal Sate East Bay and then three years law school and then the bar." Let me worry about me."

"K."

"I got to get rolling, Joe said starting up his car. I'll call you tomorrow after work."

"K, she said again. Love you."

"Love me, too" Joe responded with their old joke. He off campus.

Hitting up the Taco Bell on El Camino Real as he left Stanford, Joe drove slowly unwrapping the food as he went. All the food they had in the apartment was wine and Indian food from Naqeeb's family restaurant in Fremont. As much as the starving student in him appreciated the free food, curry every night was a little much. He turned left on University Avenue munching the burrito.

Traffic was bad on University past the train station. What did anyone expect? It was California, traffic was always bad. The weather was excellent, however. What did anyone expect? It was California, the weather was always excellent.

Three blocks up to Ramone and Joe hung a right. Polishing off the burrito, he considered his thesis presentation. Two years of hard work and the presentation was a C-plus at best. The software was a plus he thought, but his foray into the professors' privet life had cost him.

Not enough to invalidate the whole he figured.

Turning left on Channing he cruised the block and a half to the Palo Alto Housing Corporation. Five non-descript four-story buildings provided housing for 197 students in one and two bedroom units. Joe considered himself lucky. Despite the expense of the area, 1800 a month was reasonable. Heritage Park was a little sketchy, but compared to East Palo Alto located on the other side of the 101, this was paradise.

Hell, the Hewlett-Packard garage was only two blocks away. Joe wondered if apt 201E was going to be remembered the same way. He pulled into a vacant visitor space.

Angela's 2008 Lexus IS 250 was in the reserved space. Naqeeb always drives Joe thought. Predictable.

Why did they take his Camry when Angela's family always made sure she had a nice car and stuff? I would look good riding around in that he thought.

His roommate would pay the price when they got home and no other visitor parking was available. His loss.

Joe unlocked the door to the apartment and hung up his backpack. Nap time. He had been too keyed up to sleep well last night. A couple hours now would shore him up for tonight and his work shift tomorrow. It felt strange to have no classes or his project to work on anymore. Lying down on his bed he reflected that the last two years had been the best of his life. Toughest mentally of course. Even dad's thing (dying) wasn't as stressful as this. More emotional sure but not as stressful.

Being accepted into the Ph.D. program might have meant a delay in making that first billion. Buuuttt. The chaos that was 2008 made his decision easy. Stanford had come thru with scholarship and aid money as well. Added to that he had met Naqeeb and found a good friend. Really a brother if he thought about it. Weathering the Great Recession in the calm of academia was a blessing. Joe wasn't religious but the thought of it that way: A Blessing. His catholic roots were showing. Too many of his fellow baristas at Peete's were way-way-way too educated for those jobs. He dropped off to sleep wondering how Naqeeb and Angela had made out.

Two hours later Naqeeb parked his car on the street. One spot down from "the bad spot" where it had been stolen from once before. He was gratified to have gotten the car back two days later.

What always puzzled him was the strange combination of items left in the car. A baby blanket, a pacifier. A trove of religious pamphlets. As well as a host of cell phone chargers and cords. Weird. Even more so since the police had told him the car was pulled over driving slowly on the roads near downtown San Jose. The three young teenagers had bailed as soon as the cops flashed the lights after running the plates. They never came close to catching them. As he installed the club and locked the doors Angela retrieved their backpacks. "We passed, yeah!" she smiled wanly.

"Honey they were never going to spend that much time and effort and have us fail. I just wish we could have succeeded," he told her.

There was no good response from her as they trudged upstairs. Entering and hanging up their backpacks Angela noted Joe's already on the wall. "I wonder how he did," she said simply.

"I'm sure he did fine," Naqeeb said loyally.

"I'm sure he did too, I'm just asking," she said coolly.

Joe and Angela didn't fight exactly they just never really clicked.

The young man picked that moment to emerge from his bedroom and walk down the short hall.

"Well"? He demanded without preamble.

"It went fine" Angela answered.

"Fine as in Stanford bestowed Ph.D.'s and Microsoft wants to form a startup worth a billion dollars or fine as in- I hope my parents didn't turn my bedroom into a hobby room?"

Naqeeb rolled his eyes. "It went fine". The committee was very pleased. They told us afterward that we passed in confidence but that we were going to have to wait a few days before the official announcement came out."

Joe absorbed the good news. "Congrats! "How was the presentation itself?"

"Grueling. Okay" Competing answers from the couple.

"Well spill. Tell me the details," he exclaimed.

They rehashed the 6 1/2 hour dissertation defense.

"Shit," Joe said. "That's crazy. They weren't impressed you got the Kajiggers to link up?" he asked Angela.

"They were impressed, Joseph. They just made us explain every decision point and asked why we hadn't tried other methods," she said too patiently.

Naqeeb weighed in at that point to forestall the coming fight, detailing his defense of the Alternate Programming Methods section. The two men waxed technical programming talk while the evening evolved. The female materials engineer caught every third word.

"Ouch- Joe said to one of the tougher questions. That's going to leave a mark."

Naqeeb laughed. "Hey, they recognized the skills" he rejoined.

Angela suppressed a little smile. Joe could always pull him out of any funk. Not that she was jealous of him. No.

"How as your session she asked Joe seriously. And don't say Fine!"

"Not as tough as yours. I, of course, kept it real by pushing the boundaries while explaining the Predictive Solutions Matrix. All in all not bad." He finished.

"Did you pass? Naqeeb asked. Serious.

Joe hesitated. "I think so." He recounted Professor Thorson's concluding speech. I should know in a few days as well but I think so."

"You took a big risk", she tsked at Joe.

"The linking and predictive matrix are really the new things in the algorithm. Everyone is data mining. "The Matrix makes it Special" he deadpanned ala Keanu Reeves.

"Hey what's the story on dinner"? Joe suddenly exclaimed. "Let's open a crappy bottle of wine and toast our success!"

Naqeeb agreed. "We have some leftovers in the fridge from The Kitchen, he said. Joe stifled a groan. Free was free.

"Great, he said. "I'm sure you guys are going to be fielding calls all night."

In truth, all three had been texting continuously since they had sat down. They did it unconsciously.

Naqeeb snorted at Joe's comment. "Are you kidding? I had three voicemails from my mom when we got into the car to come home. I swear she got the iphone just to bug me. Thank God she can't text!"

Angela stood. "Speaking of which I have to call my parents. Then I'll be back to help with dinner." She left for her room and a little privacy.

The men assembled Chicken Tikka Masala and Buryani from the fridge containers.

"What wine goes with Indian food"? Joe asked.

"Boone's Farm Apple grape? Came the reply.

He snorted. "Since we have Merlot it will have to be Merlot."

"Why did you ask if that's all we had? Naqeeb asked exasperatedly.

"J.K." Said Joe. Oops, a text from Renee. I have to take this one," he smiled.

His friend looked at him. "Why is she rotating back to the top of the list?" He asked sardonically.

"I helped her compile some software and linked up her data tables correctly and now she wants to show her appreciation".

"You did her Senior Programming project for her in the middle of your dissertation?" Naqeeb asked incredulously.

"I just helped her over some rough patches, Joe said defensively. Now she wants to meet up for a drink."

"I thought we were watching Los Gigantes tonight," his friend whined.

"Your jealousy is cloying. Woman, let me breathe!"

Naqeeb laughed as Angela returned to the kitchen.

"What'd the 'rents have to say? Joe asked.

"They were pleased," Angela answered looking at Naqeeb.

"Pleased?" Joe said slowly. "You led a team that engineered Nanostructures and agents never before seen on earth. You almost get them to become self-aware, almost achieved A.I for the entire system. Almost become a god like figure not to mention how incredible the Nano Factory and Imaging Systems are and they think…." He wound down halting.

Naqeeb smiled at him. They might fight a little but they did respect each other's talents.

Angela smiled as well. "Joseph, pleased from my parents is jumping up and down from other parents."

Joe dished her out some rice and poured a glass of wine. "A toast then…

Doctors all are now we three,

Though not the kind our parents wanted us to be.

Sheepskins tied with ribbons and bobs,

One simple request is all we ask: Anyone got any jobs?

CHAPTER 2

May 18, 2010, 2037 Monday, Palo Alto, Ca.

The TV flashed in the entertainment center in the corner unobserved. Tuned to the Giants-Padres game in the third inning, Naqeeb and Joe had insisted on checking in to see how Timmy was pitching. So far so good. 0-0.

Not that anyone at the party was actually watching. Consequently, the campaign commercial was a doubly bizarre image out of context. Sheep with red eyes? Angela just caught that image for two to three seconds. *What? Were the sheep Cambell supports or Pullman's* she thought?

Angela was apolitical for the moment. She had too much to do. She didn't care who won as long as they were being honest. If she had to guess she thought Pullman would win the primary but she would never stand a chance against Boxer. Not in California.

She turned from the TV back to the party. A second disturbing thought flicked thru her head: This could get out of hand quickly. 33 people were jammed into the living room and kitchen of the small apartment. Talking, drinking, eating and well, partying. Not that she was against people having a good time it was just...

The party had technically been a joint idea from the three of them. Joe was as ever the social glue. Angela had no talent for organizing a social gathering. Joe was a natural at it. He had first floated the idea two days after the presentations. The three of them hadn't, two seconds of free time consecutively in what seemed like forever. To top it off word had come down from on high that they had all officially passed. Doctorates would be conferred!

"Let's just invite some friends and family and let loose a little" Joe had proposed. "We need some down time before the graduation/job/family dynamics pulls us every which way".

They all recognized that the little chrysalis of their world was about to be shattered by reality. Who knows where we will end up? That was exciting and scary and great and terrible.

Naqeeb had agreed. "We have a short window between the school storm and the job storm we should celebrate."

That settled it. Angela stood back and let Joe take over. Other than providing her guest list she just rode the wave. From her just her sister Suelyn and Harold and Betsy from the project.

Naqeeb had included the latter two on his list also. Naqeeb's sister Shireen and his brother Khushal up from UC Irvine to do a load of laundry rounded out his list.

Joe had his sister Kelly and two girlfriends invited. Any looming thoughts of cat fights were dismissed by Joe.

"It's cool" was all he would say.

The original guest list of 9 had swelled to its current couch busting level and more friends of friends were starting to arrive.

Angela briefly equated the flash mob arriving at her home with the linking of the Nano-agents she had created. Goal driven entities bent on their own agendas. Drinking all their liquor and eating all their food seemed to be the group goal. Angela snorted and thought that the newest arrivals were neighbors of some sort. People did seem to be having a good time.

She saw her sister Suelyn talking to Betsy. Khushal was surreptitiously taking a drink and eyeing Renee, Joe's ersatz girlfriend.

Naqeeb had caught the quick sip for Khushal and frowned at the underage drinking. He looked around to find Shireen to have her talk to him. He and Khushal had a strained relationship for the last two years or so. The eighteen-year-old younger brother was finding his way through the minefield of Naqeeb always did what he was told…

Shireen was plastered to Joe's side while he played bartender. Angela saw Renee and Victoria pointedly ignoring each other. Fools she thought. Shireen had exploited the opening with the ruthlessness of an attacking general. Her name might mean sweet in Persian but she could manipulate with the best of them.

As Naqeeb made his way over to her Angela caught a glimpse of Joe's Sister Kelly fending off Harold Cho. *Hi ya*!

Naqeeb slipped an arm around her and bent to speak into her ear. "I'll channel some of our less familiar guests out the door, okay?"

"Thanks. It is getting a little loud and crowded."

He swirled off into the crowd. Angela thought she had better rescue, Kelly.

Kelly's eyes had a slightly panicked look as Harold launched into a detailed technical rant about the factory. "It was the embedded tantalum wires in the lead/copper alloy encasement body that allowed the fluid in the vacuum field to rotate. The magnetic field interacted with the molecules in the suspended medium when electrical current was applied."

"What is going to happen to the Nano factory now Harold,"? Angela interrupted. Kelly sent "thank you' thru her brown eyes as Harold downshifted into contemplation.

"I'm not supposed to speak publicly about this," he started.

"Let me guess", Angela told him starting in. Stanford just reminded you that all real, physical, and intellectual property created in the Ph.D. research process belongs to the university/industry partnership right?"

Kelly who had been sliding towards her exit suddenly stopped, interested. She asked Harold about the releases they had signed.

"It's a full release. We give up all rights. Harold lamented and Angela confirmed. "That factory was my baby. I put everything into it and now I don't even own it," he wailed.

"None of the Nanostructures or agents would exist without that factory," Angela acknowledged.

"What happens to it now, Kelly asked. "It just sits on a shelf?"

Harold was silent a moment. "Stanford has asked me to join a privet enterprise they are putting together." *Ah ha.*

Angela smiled pleased for him. "Harold they know the tremendous amount of work you put in the factory. They'll be fair to you", she said into his frown of doubt.

Kelly was silent on the matter.

"It's a big deal and I'm surprised Betsy isn't joining you," Angela went on.

"She turned them down, he told the women conspiratorially. She is going to form her own startup."

"Harold! Betsy admonished joining them. "I told you that in confidence".

"That won't work for you," Kelly told Betsy seriously.

She regarded the younger woman a little haughtily. "And who are you again?" She asked down her nose.

"I'm Kelly Smithson, Joe's sister. I think you and Harold are grappling with the same problem and looking at two separate approaches." She paused for breath becoming aware of the number of eyes on her. "It's a basic question of control for work and sweat you did. Stanford owns all the research and you signed releases. Those are hard to break. Not impossible but hard. Harold created the factory and would like to replicate it for other research teams and get fabulously wealthy in the process."

Harold nodded enthusiastically.

"He can't form a startup and just copy the factory. The release undoubtedly contains a non-compete clause. Any vulture capitalist will avoid him like the plague because the university can tie you up in the courts and strangle you with paper."

Joe caught her eye as more of the party focused on the discussions. Those pre-law courses were going well.

"So Harold's approach is to join the evil empire and help build the death star." He will have a job but no real control or equity if he is just a project led or regular engineer." She leaned in to deliver the alternative. "But if he's smart he can regain equity and control in his employment contract.

Kelly had a captive audience as the Giants pushed across two runs in the 8th. "You should agree to work for them in exchange for a 25 percent equity stake or 25 percent net profit whichever is higher. Also, ask for an Officer position in the company. Chief Technology Officer sounds about right," she concluded smugly.

"They'll never go for 25 percent," Joe mused.

"Start at 25 percent and settle for 8-10. If they bang you too hard, stand up in the middle of a meeting and say: "Future Soft Technologies is at 15 percent." Kelly countered.

A soft "ohhh" came from the three people who understood the play. Harold did not.

Naqeeb clued him in. "Future Soft is a Chinese Nanotech company."

Betsy took over the explanation. "They will steal the technology and build the factories regardless of the patents or the US courts. By the time the WTO gets involved they will have booked the profits and moved on. Short of the US government going to war with China over the stuff, there is nothing they can do to them."

Another ohh went thru the crowd as Betsy blushed.

Joe stuck his two cents in with "Any US Company that manufactures in China is asking to get its tech stolen."

This led to a round of questions on how Apple was able to get away with it.

"The Apple brand is what's valuable not just the smartphone technology or even the tablet stuff", Joe suggested. "Apple could put out a blackboard and a piece of chalk and call it "iRetro" and schools would jump on it." "A quick test: Does anyone here have a Motorola phone?" Joe asked.

No one did.

"Why not? He asked. It's a better phone. Faster, cheaper, more features. Better. But none of us would be caught dead with one unless it becomes cool to own one again."

Betsy focused back on Kelly. "But why won't my startup work? She asked again. Her slight frown marred her elfin face. Betsy had the kind of blond California looks that inspired songs.

"Because they will have redress in the US courts against a US start-up. Patent law is all on their side." Kelly answered.

Betsy considered that. "But the real break thru in the imaging system is the software controls not the optics or the processor. If I own that I'm good right?"

Naqeeb perked up. Hey, I still own 10 percent of any billions right?" he joked holding up a folded piece of notebook paper with their "contract".

Kelly suffered inside. *Engineers.* "Stanford has your release as well and owns all of the imaging software in addition to the AI stuff. That contract is null and void."

"They own my stuff too, Joe confirmed.

"Yes, they do Kelly said slyly. "But what exactly do they own? What would a reasonable person say? What exactly is in your release and the non-compete clause? For what length of time is it good? Has Stanford jumped thru all the hoops it needs to in order to exercise its rights? Is it only for a narrowly defined set of technologies, or is it too broad to be enforceable? What would a reasonable person say is a reasonable amount of compensation for the work performed? Remember the Winklevoss twins and Harvard didn't get any of Facebook right? Harvard did get a nice chunk in their endowment but that's another issue". Kelly wound down.

That was a lot of information for the crowd to digest.

Betsy immediately dragged her away and started picking her brain.

The party dwindled down to a comfortable few sitting and talking. Harold looked glum considering what to do until Naqeeb took pity on him.

"Dude. Stanford has one of the finest law schools in the world. Don't you think someone over there could help you write that employment contract?"

Harold brightened considerably. *Yeah.*

Renee's Women who Rock iPod playlist was in the dock as the Giants closed out the Friars 7-6. Lincecum notched the victory.

A small cadre of family and friends sat on the couch discussing life and events.

Naqeeb loved this time. His brother and sister with him while they celebrated. Angela basking in a glow of her own. He was grateful that Suelyn was letting her shine. Joe had Kelly looking out for him. A lot was right with the world right now he reflected.

"So what's going to happen to your research Suelyn asked Angela. Her older sister was a pediatrician. "I have this vision of the government crating up the Nano-agents and storing it in a giant warehouse like the last scene of Raiders."

Angela, Harold and Naqeeb and Betsy all exchanged glances.

"That's not far from the truth," Angela said. "I'm supposed to siphon off the Nano agent fluid medium and freeze the whole environment."

Harold nodded glumly. "We have to turn over the Nano factory on Monday the 24th."

"And the agents never achieved "awareness", Suelyn asked Naqeeb.

"Nope," he replied simply.

"How did you try to make them come alive," Kelly asked interestedly.

Naqeeb sipped his beer. "Angela and I tried to mimic natural evolutionary systems in the Nano world." For instance, we set up a light source to simulate the sun. 12 hrs of light and dark to simulate days and nights. We set up a food Nano factory to

28

provide raw materials. There was another factory to tear down old non-functioning agents and constructions." He paused to sip again.

Angela nodded along with her boyfriend's recital. "Naqeeb was brilliant. He used chemical and mechanical interfaces to try to program awareness into the agents. He came up with a kind of frequency standard to mimic a computer system's input-output cycles. She grasped his hand. "He melded the human evolutionary process with basic mechanical and computer processes."

"Angela did most of the heavy lifting, he said deflecting her praise. She figured out how to align the various molecules into basic IC logic circuits and mechanical systems."

He smiled remembering the hours she spent aligning individual molecules and atoms into AND gates and Fulcrum levers and reduction gears. The Nano agents had a whole spectrum of logic circuits and mechanical tools to draw on.

Naqeeb looked to make sure Kelly was following him. She was having a little difficulty.

Joe stepped in for her. "So she gave the bugs problems and tools to overcome the problems she gave them", he told his younger sister.

"Don't call them bugs, Angela said automatically. Or kajiggers, or igits or thinga ma bobs…

"Spacely Space Sprockets they are then", Joe retorted.

"Children," Suelyn said as Kelly punched Joe in the arm.

"But they never got there huh? Too bad, Kelly said after a moment. That would have been cool."

"They linked together and changed some structures but they never lit the light, Angela said sadly.

Harold relayed to the outsiders that the final problem for the agents to overcome was linking together a set of simple circuits to make a basic light switch. That would have been a conscientious act. A willful act. There was a reward in that several Nano factories would have been available to provide more resources to the system. That would have shown them acting to improve their environment thus proving intelligence. He wound down.

Kelly stepped into the breach of lowering party morale. "Hey, life goes on right. You can always do more work right? Anyway you all passed, even my stupid brother so a toast to you all!" Glasses were raised and dutifully clinked.

Joe finally noticed Khushal chatting with Renee and wondered where Victoria went off to. Oh, well.

He leaned into Naqeeb. "Dude, keep your brother away from my girlfriend he is way to young for a car that fast."

Suelyn overheard and told Naqeeb, "While you are at it, stay away from my sister."

Angela rolled her eyes and told Shireen that she should stay away from Joe because he would never be pinned down.

Betsy joined in and told Harold that he should stay away from Kelly because she was her lawyer now.

Kelly announced that everyone should leave everyone else alone or she was going to get parents involved. That was the final word on the subject of who was with whom.

The party broke up around midnight. Angela cleaned up while Naqeeb and Joe played Xbox. Attempted to play Xbox. The HALO game kept glitching. After the third screen freeze, they gave up.

"We need to talk about the lease tomorrow" Angela reminded them. "We can extend to June 30th but that's it."

"I'm hoping the job situation will be a little clearer by then, Joe said. I do not want to be a boomerang kid".

Too many of their friends had been forced to move back in with parents. A sentiment Naqeeb and Angela wholeheartedly agreed with. "Night, you guys," he said heading off to bed.

The couple stayed up for a few more minutes talking about the party. Who'd come, who'd drank too much, who'd hooked up? They agreed that the party was a success and maybe the couple should throw more parties.

"Not beer bashes, like Joe, would throw, she told Naqeeb, but real parties. Grown up parties."

"Why does that sound like "parties after we are married? He complained.

Angela looked at him steadily. Brown eyes met brown eyes.

He grinned. "Okay, okay. Grown up parties. Dinner parties with our couple friends."

She smiled pleased. 'Can we go to bed now?"

"Yes, please!"

CHAPTER 3

MAY 19 1037 2010 Tuesday, Pleasanton Ca.

The Alameda County Fairgrounds occupied 430 acres off of Bernal road. Close to the 680 freeway, the fairgrounds were a permanent attraction in the east bay. They had been for decades. A Farmers Market joined carnival rides and a bustling midway scene. A horse racing track even ran when the ponies left Golden Gate fields in the summer.

This morning the fairgrounds were hosting a special campaign event for the California Senate Republican Primary race between Margaret Pullman, Tom Campbell,

Steve Pozner and Pete Foy. Republicans in blue California. Of course, the more rural east bay and central valley had long been outposts of conservative thinking in the liberal state. Today's candidate forum was an especially energized event for the republicans.

The Tea Party was flexing its newly formed muscles. Frightened, angry, older, uneducated white people had had enough. Anything that threatened the most privileged class of people since 18th-century English nobility was to be feared and eradicated.

The slicksters had started using the money and energy of these deluded people towards their own ends. The rally today was really supposed to be a candidate forum. Ostensibly a platform to exchange views, it had become a vehicle to stake your claim as the whitest, and right candidate in the land.

Margeret Pullman rode in the back of the specially outfitted black Escalade across the San Mateo Bridge on the 92 heading for the fairgrounds. Had she been told that the Escalade was the car of choice for black rappers and athletes, she would have steered clear. She chose it for the same reasons they did: room, ride, and prestige.

The black tinted windows hiding her from the prying eyes outside. She spoke into the phone with a snarl on her face.

"Andre let me say this again. 38 million in bearer bonds ready for Marcus at noon tomorrow." "I don't care, she spat in response to a statement on the other end. "You are already bleeding me. Fail and your whole family will regret it." She hung up.

Blood suffused her face. "God damn Dodd-Frank she exploded. The three others in the SUV waited out the storm. "Penny Pullman's little girl is not going to be cheated by some Swiss banker…"

She turned to address the man on her right. "Marcus if the son of a bitch even mentions signing for the account close out, you call me."

"Right away, sweetheart. Don't worry its all set. George is going to take me to the jet right after the event," He swept a hand out towards the driver.

A moue of disapproval pursed her lips. "We need this money," she said to no one in particular.

The other man in the car in the opposite seat, James Cantrell, smoothly folded in. "Andre will play ball. The personal account holds over 45 million. That's 7 million to ensure no SARBOX foreign Swiss bank account- IRS problems," he said.

Cantrell used the acronym for the Sarbane Oxley, act which made reporting foreign bank accounts to the IRS mandatory.

That law and the Dodd Frank act, passed during the height of the economic collapse had meant shifting money around was now tricky. Very tricky, but still a priority.

Pullman needed to launder the money into the firm's accounts and then pass it over to her personal accounts as quietly as possible.

There was a general silence to his statements. The mood in the car suddenly tense. They were walking a tightrope and these three knew it. "Let's review the event", Peg directed, refocusing them.

"Okay," said Jimmy. "The candidate's forum is where we want to hammer our major themes: 1) The US is screwed up right now and people are hurting. 2) It's Barak Obama fault that Wall Street got bailed out and now he is taking away people's health care to give it to illegal aliens thru Obamacare. 3) We need to take our country back!" He paused drawing breath.

"Make sure you set the theme and then hit the supporting statistics hard!" Then give people time to react after you say the lines."

He demonstrated: "The America I know is going in the wrong direction and it's affecting people's lives. 8.7 million Jobs lost to the Democrat recession. Pause- 19 million people unemployed or underemployed. Pause. Wall Street bailed out to the tune of billions while Main Street suffers. Pause."

Cantrell looked at her. "Do you have that cadence?" She did.

Cantrell's phone beeped. He read the text. "MSNBC is calling Campbell's ad from last night "Demon Sheep". "You have to work that in", he said seriously.

Peg considered this while Marcus made her another drink. "Will the crowd know what I'm talking about'"? She asked.

Cantrell grinned. You bet! Armey's reps will be running thru the crowd priming them. They'll be ready!"
He was referring to former Texas congressman Dick Armey's Freedom Works representatives. Armey and the Koch brothers smelled change in the winds with the Tea Party movement. Recognizing the vacuum of anger and hopelessness they had co-opted the small grassroots organizations into an extremist money making conservative machine. With the emphases on money making.

Margaret Pullman recognized the synergy between Freedom Works and her own political ambitions right away. A built-in vote generating base was assured. All it took was a check. The Democrats and establishment Republicans would never know what hit them.

She consulted with Cantrell, her campaign manager, about the demon sheep lines as he pulled the highball from her hand and swapped it with a bottle of water. High 80's in Pleasanton today. Hydration was the key. That and being sober for the event.

As the SUV pulled into the fairgrounds final preps were made. Clothes, hair, makeup checked. Nice but not too high end. A conservative pants suit for her while her husband Marcus was in a stylish gray suit. It wouldn't due to be seen in designer clothes right now. Luckily she was fit and trim, even if she drank too much. 57 years old with styled hair and prominent cheekbones, she still looked attractive. Just not to her husband.

Eggert met the car as they parked near the other two advanced team SUV's. He spoke with Cantrell who was the first out the door while she and Marcus waited inside. Eggert, their head of security relayed orders from his walkie talkie and he moved off into the fairgrounds with the two white members of the four-man security team. *Athos and D'Artagnion*, Peg thought.

The black male and the Latino female security members stayed with her and the vehicles for appearances sake. Porthos and Artimus she vaguely remembered. Eggert loved his Dumas. She knew that much anyway.

The applause from the small crowd as she emerged from the car was muted. Attracted by the SUV's and the activity, they quickly realized it was just a politician they were seeing. They wanted to resume their own activities. Peg waved and smiled gamely as she waded out to meet the people. She shook hands and shouted greetings to startled fair goers. It took them some minutes to make way to the platform. The crowd swelled as they approached.

The Musketeers kept the uniformly older, white crowd at bay while she walked up the steps. She joined the small group of white men on the stage. The only other candidate she shook hands with was Pete Foy. She wasn't going to give Campbell the legitimacy by shaking hands with him. Foy was no threat, however.

By agreement, each had 15 min to address this group of undecided voters. Getting 4500 people out on a weekday to hear primary candidates speak- spoke volumes about her teams' efforts today. Pullman supporters dominated the crowd. Several hundred had the FastFac t-shirts emblazoned with the FF design that had become part of the campaign.

Peg Pullman spoke after Insurance Commissioner Steve Pozner. Fifteen minutes of talk from a mainstream republican who favored pragmatic compromise solutions to California's problems left the crowd restive.

Peg rose to her feet and sipped from her water bottle as she surveyed the crowd.

"Good Morning. Thank you for loving California enough to come out to listen to politicians speak in this heat." Small chuckles sounded in the crowd. "Sorry, I was a little late this morning. The San Mateo was backed up and there was no *way* I was going across the Bay Bridge!" More chuckles. "There's a typical Washington D.C. Jerry Brown/Barbara Boxer/Nancy Pelosi type of project for ya!" The responding roar was visceral. "The Bay Bridge replacement! Billions over budget! Decades late and still not ready!" she thundered.

She paused to hear the response; let them get into it. "Designed by fools, built by idiots, watched over by incompetents and paid for by your hard earned tax dollars!" She let the applause wash over her.

She started pounding the themes they had practiced in the car: Theme, attack line, pause. Theme, attack line, pause. She felt the rhythm and let the crowd sweep her along. She struck gold when she called Democrats sheep. "Not demon sheep mind you but regular uninformed sheep!" The BAA's that came back from the crowd brought a smile to her face. She kept pounding away. Pausing to let the answering roar come back at her she paused again. The approval roar followed by a moment of silence. A lull in the pace.

A male voice shouted up from the crowd, "Fuck Obama, he's an n_____."

The loud reaction was mixed. Gasps of shock. Cries of approval. Shouts of disapproval. Peg had a McCain moment.

During the 2008 Presidential campaign race, Republican nominee John Mc Cain the Senator from Arizona was taking questions from the audience. Normally a time killing device to show how much you listen to the people when a confused looking elderly woman posed a blatantly racist question to him. Civility won out and McCain told her that no the Senator from Illinois was a good man who loved his country.

Peg was not about to make that mistake. She plowed ahead as if she had heard nothing. "I'm going to need all your support and efforts if we are going to take back our country from Obama and Boxer! God bless you, God bless California and God bless the USA! She left the stage to organized chants of "Run Peg Run!"

She quickly retreated to the SUV not even listening to Campbell and Foy. Polls had her at 35% and climbing. No one else was even at 16%. She thought she needed to start concentrating on the general election. The June primary was in the bag.

The cool of the air conditioning was a relief after the heat of the stage. She was much better in small crowds' vs the larger events but she felt she had pulled this off. Get her one on one with people and she always closed the deal. That's how she conducted business.

Cantrell climbed into the SUV and removed the drink from her hand. "We still have the speech at the Jewish Federation tonight" he reminded her.

"How many times do I have to say I'm a friend to Israel, "She whined.

"Every vote and dollar gained are taken directly from Boxer," Cantrell told her flatly.

He waited a beat. "Did you hear the racial slur?"

"Of course I heard it, how could I not?" Venom dripped. "God those people are idiots!"

Cantrell was silent. Then he said, "Marcus is out talking with reporters. I just left McClatchy. They heard it. They want a response."

"What do you think?"

Cantrell shrugged. 24 years in business with her and the last 13 months as her campaign manager had taught him her staggering indifference to everyone but herself. "We double speak them. We didn't hear it. We don't think that's what was said. Racism has no place in 21st century America, blah blah, blah."

They left the SUV to give the newspapers and TV what they wanted. She would be featured on the 6 and 11 o'clock news bemoaning the fact that she had been the victim of a) Horrible attack ads, b) a manmade glass ceiling and c) a totally unfair double standard when it came to women in politics.

Later outside the SUV, Cantrell huddled with staffers and Eggert as the schedule was adjusted and updated. Eggert and the Musketeers went to the downtown SF hotel where the speech was occurring. The J Fed speech went off at 7:00. She had an hour worth of meet and greet before with a half hour after. That should dovetail with the newspaper deadlines to make the morning paper and the TV could go with coverage of the morning's events and the speech. Always work to their schedule was Cantrell's motto. Marcus the driver and George headed to the private terminal at SFO where the corporate jet waited to fly him to Zurich.

Cantrell and Peg had business to attend to in Mountain View.

FastFac headquarters was kind of disappointing. It looked like an office building. It certainly looked nothing like Elison's ode to smoked glass that Oracle built up the 101.

And that space ship Jobs was proposing for Apple was a little gaudy. Rumor said he was sick. Peg hoped he would die soon.

Margaret Pullman's dirty little secret was that she was in trouble. Although she was beginning to suspect that the secret was not so little and not so well, secret.

FastFac was sinking. Where they had once dominated their segment, people now turned away in droves from the limited use of the e-mail messaging system. She had failed to see the smartphone and tablet revolution coming. There was a newer sexier version of the messaging device hitting the streets on 1Jun. It had better work.

This afternoon was a major status report to the board. Corporate boards tend to get cranky when 60% of the market value disappears from the firm in five years.

Margaret Pullman was not immune from the losses. Her personal fortune shank from 17.2 billion to 6.8. 2008 was horrific for both the business and her own finances. Even worse the board was sniffing around some of the accounting tricks Cantrell and she used to paper over the losses. If they ever were forced to admit to the true value of some of the acquisitions, they were truly fucked.

The boardroom contained the required gravitas for its station. The table screamed this was where billion dollar deals get done! Three big screens dominated the inside wall tuned to FOX.CNBC, and Bloomberg. The glass windows normally showed the bay and the huge blimp hangers at Moffett field. Right now they were opaque to allow the power point presentations.

Cantrell and Peg entered from the hallway halfway thru the meeting. As they sat a board member named Anderson was pestering the marketing VP. "How can you predict 5 million units in the first 3 months he asked tartly? "That would double our current user base."

"That's still just half of what our user base was in 2006" the VP countered. "The new cloud features will siphon off Apple users and bring back former users, he countered. He squared to face the board at large. "We still have the government users, he reminded them. Even the President uses the #2212. The new Scimitar will draw in 800,000 government users. The civilian contractors for the DOD and homeland will have to follow suit." He let that sink in. Silence hung in the room. *That could work. That could be true... That might happen if...*

Peg seized her moment. She understood the room and the want to believe in what they were doing. Standing, she drew eyes. "Ladies and gentlemen. Mike Morris is right. This is the perfect confluence of time and technology. The Scimitar is priced right, it's a huge upgrade in performance and we have the support in place to satisfy our customers."

She noted the similarities between this and her stump speeches. "What we need to do is focus on the roll out. 1 Jun will be a monumental day around here! We need to build buzz. We need to build excitement! I want to see customers standing in line for our product! Media coverage. We have to get the employees involved. Scimitar is poised to lead this company back to its rightful place as the leader of Silicon Valley innovation! Let's make this happen." She concluded to applause from her own board. A little belief goes a long ways she thought.

Of course, they went along. What other choice did they have? If this went south all their stock options were worth pennies on the dollar. Besides most couldn't trade the stock now anyways. Insider rules had them locked out for the next three months. Anyone who didn't have a preprogrammed set of trades already set up was screwed.

Peg did of course. As the majority shareholder, she had been quietly selling for four years. *You would think 753 million dollars over four years would be enough,* she thought bitterly. No such luck. Her costs were enormous. Both the business and personally her finances had a thousand leaks big and small. The houses were money pits.

The Hillsborough estate was bad enough but the vineyard and the New York apartment were prohibitive. Peg recalled a joke Francis Ford Coppola told her one time in Napa: "How do you make a small fortune in the wine business? Start with a large fortune and buy a winery." Ha Ha. Her estate made a nice wine. That nice wine in no way justified the 4 million a year it cost to make. She could sell but she couldn't get what she paid for it yet. The market needed to bounce back more than it had to make her whole.

In short, she was hemorrhaging money even before she decided politics was calling her. Fifty million seemed like a small price to pay to be a Senator. Scimitar had to work. It had to buy her time. There was nothing in the hopper to save them if this failed.

Cantrell followed her out of the board room up to her office. She poured herself a drink. Cantrell let this one go. "Jimmy, what do we do if Scimitar fails"? She asked him and herself.

"Nothing," he said quietly. It'll be over by then"

He perked up. "We just need it to be successful enough to buy us a year. "In July you can resign as CEO to avoid any conflicts of interest in the general election. You can sell unrestricted 90 days after that. It's going to take 6 to 9 months to unload the bulk of your holdings." The price will go down but we can cover that by announcing that you are just diversifying your holdings."

He reassured her. "The 1 Jun roll out date is a genius. We cover up the dismal Q1 numbers with the product announcement. We don't have to report any hard and fast numbers on Scimitar to the street until Sept. We plant rumors and leaked stories of spot shortages. We get the new model in the President's hands and let him be seen with it. Once we win the election, they can try to hit us if anything goes wrong with FastFac but then it just looks like sour grapes." He laid it out for her beautifully. Again a little confidence went a long way. Cantrell moved to massage her shoulders. Peg allowed it grudgingly.

"When do we have the analyst call?"

"Not till 31 May to preannounce Q1 and hit the Scimitar." He moved to kiss her.

She pulled away. "Not now". Come over tonight after the speech."

Jimmy nodded. He didn't feel guilty. Peg and Marcus had not been friendly for decades. Besides Marcus had George. If Peg wanted a little comfort he would provide it.

Cantrell wanted the gravy train to continue. CEO, Senator, Ambassador. Whatever Margaret Pullman ended up doing he intended to be right next to her. He locked up her liquor cabinet and told her to get some sleep. He had hours of calls to make and would be back at 4:30 for a briefing and a round of campaign calls.

Peg groaned.

"I know you hate them but we have to reach out to the average citizen to have you connect. They'll rebel if they think you're not one of them", he concluded.

Lying on the couch she said "Campbell is right. They are all sheep." Cantrell shut the door.

CHAPTER 4

May 21 16:32 2010 Thursday, Palo Alto Ca.

Angela hung the backpacks as Naqeeb set the box on the kitchen table. Neither spoke. Angela busied herself setting up power cords and laptops on the table. A power strip was plugged into the wall. Naqeeb removed the oven-toaster sized object from the box and carefully set it down.

Looking at her he started, "this is a bad idea...

"I'm not having this discussion again," she said sharply. "I know the risks and you agreed. "What's done is done."

Naqeeb drew breath and stopped himself from saying more.

"6 a.m. Monday we meet Mike and Harold at the HAZMAT area door and they siphon off the fluid and disassemble the factory. We head into the Nano fab area and start the lab pack out. We turn in the laptops and that's it. They think it's sitting in the HAZMAT locker right now. Mike's fine with this. So is Harold". Angela concluded flatly.

She fiddled with the factory's connections cables. "I want some more time with the agents in the full field environment. We have to freeze them on Monday. It's a waste to let it sit idle all weekend like that!" She said slyly, "I know you want to work on this some more..."

Naqeeb considered. He had agreed initially but was getting cold feet now. He just thought that they were not nearly as slick as she seemed to think. However, he did have some new patches he had worked on and now he needed to merge the versions. He also wanted to copy the final three programs, A.I. Command and Control and Imaging into his own computer. Stanford owned it and he couldn't sell it but they couldn't keep him from fiddling with it.

He looked at Angela thoughtfully as she brought the imaging system up and ran the software. Her university computer had the imaging program while his had the C&C and the A.I.

The second port on the factory was connected to his own laptop to merge the versions of the software. All in he thought there was a 5% chance they got caught. The robbery itself was low risk. They had a property pass for the factory and laptops but no one even challenged them. Harold had carried it out to the car with them.

There were risks. Not just real physical risks, as the factory, was radioactive, but others as well. The amount of material was quite small but still detectable. The main alpha and beta particles were contained in the lead body of the containment vessel. Even the pure water medium provided some shielding. No, the main risk was to their careers. Not glow in the dark radioactive but too hot to employ anywhere. He didn't think the F.B.I was going to bust in. Not really. Maybe.

Angela glanced at him again. "No the F.B. I. is not going to arrest us."

"Stop using my brain," He told her using a private phrase. "I am way too pretty to go to the big house," he told her brushing a hand over his own smooth cheeks.

"Yes, you are" she agreed. And then she kissed him.

"Dammit", was all he said and then started into work on his laptop.

This reminded Angela of many long nights in the Nano building. She and Naqeeb met at the initial project brief when they both signed on to the project. During the subsequent weeks, it became clear that she would be the materials and Nano lead while he would head the small team of programmers. He was tall, good looking and smart. 24 years old with dark hair and soulful eyes, he radiated a strength. He was also kind and funny and honest. And Indian.

Angela felt a huge stab of jealousy when she saw Betsy flirting with him over the imaging specs. He had gently put Betsy off.

For 6 months it was 18 hour days and nights constructing the factory, writing software, making the agents. Bugs Joe came to call them. She blushed remembering how it happened. She and Naqeeb were aligning carbon atoms into an OR GATE configuration when she looked at him and blurted out "What is that smell?"

Naqeeb blushed to the roots of his dark hair. "I had to pull a shift at my family's restaurant," he said defensively. "I was making naan most of the day".

Her turn to blush. "Oh, I didn't know… It smells good… stammering. "Like bread… I didn't…"

Naqeeb leaned in and kissed her. Just a straight forward "How the hell are you", kiss.

When they broke apart he seemed a little confused. "Sorry, I needed to do that. I know we probably shouldn't date as co-workers but I'd like to get to know you." He wound down not good at the social interaction.

"Get to know me? We spend hours and hours together every day and you need to get to know me?" She said coyly. She would NOT flip her hair!

"We spend time together but we don't know each other. You didn't know I had a job", he pointed out.

"I see, yes. But I don't date."

The dance commenced. It might have gone smoother if either one had more experience. Cultural mores, family values, and school/work schedules added up to the fact that neither one had dated much. Or was good at relationships. But they learned. A simple reality drove them back together time and time again. They liked one another. They both found pleasure in one another's company. When the respective parental forbiddance came, they fought it. Not the drama of Romeo and Juliette. The simple determination of two people in love.

Even Naqeeb's best friend and roommate was going to have to accommodate Angela. The work and their relationship progressed. It evolved.

Now they were committing grand theft Nano together. As they finished setups, Angela leaned in and kissed him again. "Sorry I just needed to do that" she echoed.

"Don't think your womanly ways are going to get you out of this," he told her. You'll regret jail…"

Hi, Ya! "Just compile the new software patches please!"

Grumbling he connected his personal computer to the second I/O port on the factory. He started into work.

The gray colored basic box of the factory was about 24 by 30 by 12 inches. It had very little outward markings or features. The ceramic outer case had the two computer I/O ports lined with a rubber gasket to hold the USB female connectors in place. The simple A/C power connection looked like any other computer power port. One lone LED sat on the upper left "front" face. Front for lack of any other definition. The top of the case contained a circular aluminum lid inset into the body. It was held in place by twelve screws. About the size of a frizzbe disc, the lid was just an access plate to the inner body.

That inner body was the lead/copper alloy encasing the vacuum field and the fluid medium. The containment body was raised off the bottom of the ceramic vessel by three pegs about 3 inches high. Underneath the containment body were three electronics packages connected to the containment vessel. A power conditioning battery unit with cables running outside to the other units and to the outlet was the first package. The Vac Ion unit pump and the imaging electronics unit were the other two.

The front wall held the IO unit over the connector port. The containment body also had an access lid. This one clear glass about the size of a coaster. The center of the lid contained a vacuum trap port which allowed a tube to extend from the top of the suspended fluid body clear outside the lid.

The trap tube allowed for the introduction of materials without losing the whole vacuum field. The inner fluid medium did deform when the trap was opened to allow the material inside, but the pump adjusted and the water regained its spherical shape quickly. It played hell with the imager but that couldn't be helped. That inner fluid medium was just pure water. De-Ionized pure H2O. 10 moles worth. About 180 grams or 6 fluid ounces.

With the Vac pump running, the water held a perfect sphere in the vacuum, suspended about a ½ centimeter from the vessel floor. The vac pump and the Ion circuitry ran an electromagnetic field thru embedded tantalum wires in the fluid that provided the suspension circuitry.

The whole setup was quite stable from an outside perspective. The water didn't have to do anything just sit there and hold the Nano-agents. Visually the ball of water was just that, a ball of water with thin wires running thru it. Two micro thin wafers about 1 cent square adhered to the surface of the water thru atomic level surface tension. For all the world they looked like those Listerene dissolvable strips that freshened your breath.

Gold wires ran from the silicon wafers back down to the imaging unit. The second wafer and its wires ran back to the IO unit on the wall.

Two other items adhered to the surface of the water but they were invisible to the naked eye: the Nano-factories and the evolutionary devices Naqeeb and Angela had designed. The tiny sun which emitted light 12 hours a day. The cesium frequency standard that mimicked a heartbeat like once a second timing signal. The other structures inside the ball were the food machine providing raw materials and the recycling agent they thought up, which broke down the agents.

The 10-micron thick gold wires running from the wafers to the units had one major drawback, agonizingly slow data transfer. The imaging system had to have its wafer twisted and flexed to focus the optics portion. It was taking hours to get the imager focused and the resolution adjusted.

They feared that the IO command and control unit would have a similar data transfer rate. If they ever got to use it.

Naqeeb completed the bootup and merge from his computer. "We have about two hours before the new octal based chemical sequences need to be injected," he told her.

"Okay, she replied, let's see what the agents are up to."

Angela had the imaging system at the lowest resolution possible looking for the larger known Nano factories. The larger structures of the food unit and the recycling center were oriented across the Nano universe from the imaging wafer. Each one of the factory units was comprised of hundreds of millions of carbon atoms linked together. Around this basic framework were millions of other atoms fashioned together to provide the framework of the agents. Raw materials were available in abundance in the medium. Gold, silica, carbon, as well as technetium, rubidium, and polonium, were all available. The latter three were the radioactive elements.

The other mini factory contained rotating groups of atoms that acted like crunching wheels to tear downlinked components. Carbon tubes one atom in diameter moved the material over to be sorted and provide an input to the other factory. Each side represented a triumph for Angela and her team. These were the second functioning Nano-factories ever built.

It took her an hour to get the factories in focus. The first good pictures were resolving themselves on the laptop when Naqeeb dumped in the Silicon Dioxide.

They discussed the factories and whether they were any identifiable changes in the structures. There didn't seem to be any changes. Naqeeb complained that it was hard to know if they were seeing the same three D image from the same angle in each of the series Angela had stored on the laptop.

"It's kind of like looking at aircraft silhouettes from WW II", he said. "Note for next time: Put identifiable markers into the factory bodies to allow better orientation and recognition."

Angela made a notation in the notes.

"Make sure you shift those notes to my computer before we turn these in. Someone may notice an entry outside the lab hours."

She agreed. "Now let's look for the agents", she told him.

Angela adjusted the focus and pulled back on the resolution. She ohhed with excitement as the laptop clearly showed an agent with two larger appendages linked to its body. Angela stored the image. They only had two other photos like this. She and Naqeeb discussed what function the tubular carbon pipe might serve.

Joe returned home to find them debating heatedly. He stopped dead seeing the Nano factory on the table. He looked at Angela and Naqeeb while they stared back.

"All right! Are we stealing it or just playing with it?"

"I knew you'd take her side", Naqeeb grumbled.

"Angela, this was your idea?" Incredulous. Surprised. "I'm so proud of you!" Joe enthused.

Reigning him in she said, "We only have this for 56 hours. We are due to give it back to Harold at 6:00 am Monday." She warned him with a look.

The newcomer sat at the table and said eagerly: "Can you see the agents?"

They brought him up to speed on the structures and the attached parts. Joe whistled at the picture of the Nano agent.

"How many of these agents are there you think?" he asked her. Serious face.

She considered carefully. "That's tough to know exactly. There are just over 6 fluid ounces of water and 1 mole of Carbon 12." She told him.

Joe interrupted quickly. "What's a mole again and tell me without putting me thru a chemistry class?" he pleaded.

Naqeeb smiled.

"A mole is based on Avogadro's Constant. 6.02 times 10 the 23rd power molecules in one mole of a substance gives you its atomic weight. Carbon 12 is actually the base unit of a mole. It's also a very flexible, stable base element for the structures. Carbon is the building block of life." She told him.

"You don't have to sell me on carbon, Joe said airily, I'm a big fan."

"So I have 6.02 times 10 to the 23rd molecules of carbon atoms in there." Basically, 60 followed by 23 zeros molecules."

Joe's head spun. "It's like Peta flops in computers", he said referring to the number of times a logic circuits changes states.

"Exactly" confirmed Naqeeb.

"But this guy is not just one molecule of carbon," he said pointing to the screen. It looks like millions and millions from the pictures," Joe said. "You never sat and arranged all those molecules did you?" He asked Angela doubtfully.

"You are right. I did not. I made the factories that arranged the atoms into basic shapes. Huge flat sheets. Basic boxes and huge pyramids. Tubes like that" she pointed. "Even spherical fullerenes," she said proudly.

"Bucky balls?" Joe said.

"I'm impressed Joseph, Yes Bucky balls. Carbon is wonderfully malleable at the Nano level. It also lends itself to linking and that's what you see here. The agents have linked together into a massive structure. There are over 100 billion atoms in the agent *body* if you will."

"Not just carbon though," Naqeeb prompted her.

"No, No, other basic elements too. Silica, gold, technetium, radium, and Polonium, also Rubidium. All those elements are stable crystalline structures that lend to linking as well"

Joe absorbed this thoughtfully. "So this box and system are radioactive?"

"Yes slightly. The amounts are small but detectable. The materials and proportions of the basic elements react well to the chemical and electronic programming that Naqeeb put in." The man bowed at the mention of his name.

The three sat silent for a moment. "I'd' say there were 100 trillion of the agents. Perhaps a quadrillion of the smaller bodies." Angela concluded.

"Look there," pointing at her screen.

The imaging system had pulled back so the individual agent appeared as a dark point on the screen. Several hundred dark points scattered across the viewing area.

Joe and Naqeeb sucked in a breath. Those points were the other agents. Joe get us some drinks while I adjust the system".

Joe returned with the requisite Redbull energy drinks and Angela fiddled. She pulled back one level and the picture resolved to show six large agents surrounded by hundreds of smaller bodies.

"What are those- more agents"? Naqeeb asked.

"No, those are the smaller bodies I mentioned". "We just started seeing these when the project wrapped up." "These don't have the same large carbon bodies or the interspersed Polonium or technium molecules like the others," Angela said half to herself.

"Wait. Isn't poloniums what the Russians used to kill that guy in London?" Joe asked starting back from the box.

"Yes Joseph, Polonium 210 is poisonous." Don't drink this water", Angela advised.

They waited while the imager processed one of the larger bodies into a clearer close up.

"If you didn't design them then why are the smaller ones different from the larger ones?" Joe asked.

"I don't know" frustration showed in her voice.

22 minutes dragged by waiting for the resolve.

"This is like dial up porn," Joe quipped.

"We only have 2 more hours of daylight in the system before the sun shuts off." Naqeeb reminded them.

As the large agent's body was revealed at an atomic level Joe kept asking what the internal structures did.

"We don't know. That's what we were debating when you came in" Naqeeb said irritably.

The next two hours produced one more picture and a series of circular arguments about what was what. They shut the imaging system down while Joe printed all six pictures of the large agents. He groaned as he stretched while retrieving the prints. Had they been sitting for almost four hours? Naqeeb and Angela even longer.

It was fully dark at 8 o'clock at night outside. The complex was remarkably quiet.

Spreading the six pictures on the coffee table they sat on the couch. Joe had to scoot Catpuchino' Angela's cat out of his spot. He took in the prints.

"This is fucking impressive," he told them. Angela blushed. "No one in the world has achieved this. Those committee guys should have carried you out on their shoulders."

"I tried to tell her," Naqeeb started.

Angela cut him off. "Let's focus! "What does this look like"? She commanded.

"It looks like an x-ray of those kennect toys," Joe joked.

"That's exactly how we should think of these. An x-ray of a human or a computer." Excited she rose to her feet. "What would a human x-ray show?"

"Bones, heart, lungs organs," Naqeeb supplied.

"What about a computer?" she asked the men.

"I don't know. CPU, power supply, motherboard, I guess, stuff like that"

"I told you that before," Naqeeb said. Angela started in on the same debate.

"Maybe it's a hybrid of both" Joe said forestalling the fight.

They all contemplated that. The three spent the next hours anthropomorphizing the agents. "That could be a mouth." "Those look like arms" "A CPU?"

Angela stopped. "The real problem is that we have no map, no references," she concluded bitterly.

Sitting back, Naqeeb blew out a breath. "We are approaching this the wrong way. How did the renaissance doctors figure out human anatomy?"

"They cut open dead guys" Joe answered.

"One?"

"No, lots of them!"

"Form follows function right?" Naqeeb said.

Joe caught his excitement and fed off his idea. "Man, you're right. We need more pictures. Many more. Then we can start looking at similarities and dissimilarities amongst both the large and small agents."

"If we see the same kinds of structures oriented the same way we can then draw conclusions about what they might do."

Agreeing, Angela added the negative. "And if we see differences we can draw those conclusions as well." She was excited now too.

"So we need more pictures" he concluded again. "Is there any way the imaging system can be set to automatically take pictures as the agents come into view?"

"No, someone has to be adjusting the viewing wafer to get the focus correct," Naqeeb informed him.

"Man, what idiot wrote that software? A decent programmer would put in the flexibility to..."

Naqeeb lunged for him as Joe lurched off the couch smiling.

"Boys!" Angela scolded. "We need to man the viewer starting at 8:00 am tomorrow. We should get 30 or so pictures based on the 20 min it was taking to focus in on the agents at the whole body resolution. We could get 30 more on Sunday, she theorized. "I'm available both days." Naqeeb"?

He had the lunch shift. "I'll be off at 4:00."

"I'm back on at Peetes at 3:00" Joe put in. I'll help in the morning assuming I can get up that early. He went to the kitchen. "What kind of idiot designs a light switch that only runs from 8-8 without being able to turn it on at will?" Teasing.

"It's called circadian rhythms, douschbag," Naqeeb retorted.

"Why you got to get locusts involved," Joe asked innocently.

"Only you would design a sun that you could turn on in the middle of the night." "Can you imagine how fucked up Neanderthals would be if someone blazed on the sun at night?"

"I'm the Neanderthal in this scenario?" Joe laughed.

"If the protruding suborbital ridge fits...Naqeeb said.

Angela collected the pictures while the boys debated the relative merits of cavemen.

"Who won the Giants game?" she asked to get them off the subject.

Crap. Zito took the loss in Arizona 7-8. Crap. They were 22-18. Maybe this wasn't their year. Angela let them watch the Sports center. She tolerated baseball because Naqeeb loved it. Shireen told her she should develop an interest that would drive Naqeeb crazy. Opera for instance. No, that would be cruel. They left the Nano factory on the table hooked up and ready to go. It was not the first time they shifted meals to the living room when work dominated the kitchen table.

That weekend were days Angela would look back on fondly. Except for some breaks the three worked together continuously to take pictures of the Nano-agents and other structures. They got 71 detailed pictures of the larger agents and 12 clear shots of the smaller bots. Bots being a Joe term.

They got a little lucky Saturday. They had a three-hour pass in review as Naqeeb called it.

12 agents and 6 bots drifted by the imaging wafer at the right distance. Every 10 min the data stream resolved to show clear pictures. They happily snapped away. Naqeeb left his programming computer hooked up if they wanted to try something but nothing occurred to them.

Sunday night at 7:47 Angela shut down the imaging system and disconnected the computer. She removed the data to her own notes and computer. Joe wiped the log history and other system files to remove their digital fingerprints for this period.

"It won't hold up to a full forensic search but a cursory look will be fooled." "I'd have to wipe off things to completely remove our activities."

Naqeeb agreed. "That should be enough".

They gathered again to be surrounded by the efforts of their labors. The pictures spread out around the living room. It took an hour or so for the pattern to reveal itself.

The larger agents could be group loosely into three categories: A basic large box, a smaller box with various appendages attached and a mid-sized irregularly shaped cluster of atoms. The twelve pictures of the smaller agents revealed no patterns. They set those aside concentrating on the larger agents. Naqeeb laid out the 22 pictures of the large square shaped boxes in a 4 by 5 grid with the last two flanking the group.

Looking at Angela he asked, "Honey what you can tell us about the molecular makeup of these agents"

She studied the shots. "They all have a large carbon lattice cube embedded with rubidium, technetium and polonium molecules. There are regularly spaced gold and silicon molecules also embedded along the whole cube," she told them.

"And it looks like they linked together like that from the smaller units and the available elements," Joe stated.

"Yes, they did."

"And that's not signs of life?" Or intelligence?" He asked. "Growing like that?"

"Not necessarily no. It's like growing rock crystals from those kits as a kid".

"Maybe your childhood," Joe muttered under his breath.

"Yes, mine. But the point is the crystals grow because the conditions are right not because they are alive or intelligent." "That's what we set up here." She eyed Joe as she concluded.

Naqeeb had been looking at the pictures with a frown. "Joe look at this- No. Let's try something instead." Close your eyes," he directed. "Clear your mind." Keep your eyes closed. When I tell you to open them, look at the pictures and tell me the first thing that pops into your head, Okay"? A nod from Joe. "Ready". Another nod.

"Now."

Joe focused on the pictures and blurted "SIMM chips."

Angela gasped.

"The way you had them grouped reminded me of the layout of SIMM memory chips on a motherboard". Joe looked at them both. Memory?

The thought hung. Form follows function.

"Let's look at the other two groups," Angela directed. They quickly assembled the 28 shots of the smaller agents with the appendages. Various items jutted from the main bodies but two stood out as being almost universal. A tubular pipe and tee shaped item that looked like an old TV aerial.

"Could those be picking up radio waves? Joe asked excitedly.

"I don't think so", Angela told him bringing him down a bit.

"They could be input/out devices," Naqeeb said.

"That's it, that's it, that's it" Joe chanted.

"We only think that's it," she cautioned. "We may be projecting what we want to see."

"It's the best working hypothesis we have," Naqeeb countered.

In any basic computer system, you need three things: Memory, an input-output device, and a CPU. They quickly assembled the last group of pictures. The assorting produced 21 pictures of the mid-sized irregularly shaped agents.
The implications of what they might be seeing washed over them. All three turned to look at the Nano factory sitting placidly on the table. The LED was definitely NOT lit. That brought a little sigh from Angela.

"Why didn't they light the light?" Joe asked dejectedly.

"That's a pretty big leap," Naqeeb said. "Maybe we are just seeing natural types of growth based on the available elements" he cautioned.

"That makes sense, Joe said to Angela's nod "but it feels like we were on to something."

"Could the Nano-agents open the factories you built as the reward without lighting the light? Joe asked her.

The couple huddled discussing the question. "They could" seemed to be the consensus.

"I didn't think about making it impossible for the functions to be accessed without passing the test." Angela was silent. "It's like a drawbridge on a castle. When it's raised no armies could get in. But it wasn't meant to keep out air. They could get in there. If they needed to", she concluded.

"Necessity is the mother of invention, Naqeeb quoted. Dang! And we wouldn't know a thing about it if they got thru." He stretched. It was 11:30 and they had been at it hard the last few days.

"Can we get a picture of the factories to see if they working? That will let us know the castle walls have been breached." Joe asked casting about for a way to KNOW.

"Too late. The sun is off and we have to have this back at 6:00 tomorrow, or today in 29 minutes," she reminded him. Downcast she vented, "Hi ya! If we had made the connection three hours earlier we could have focused on the factories and proven it one way or the other." She let up.

"You guys have to go to Stanford and get them to give you more time", Joe started.

"We tried that the day after the party," she told him. No luck. Stanford if freezing the research right here."

"Literally" Naqeeb put in.

"That's why we took the factory this weekend. We knew this was going to be the last shot."

Shit shit shit. They debated scenarios back and forth for another hour. Joe was very firmly in the *screw it lets look at the factory* camp. "Christ this is Nobel-level stuff here guys"

"It's jail time if we admit to stealing the factory. Not just us but Harold and Mike too." "No way have we screwed them over," Naqeeb said with finality.

"Crap and you can't go somewhere else and replicate this can you?"

She faced Joe and said, "No I cannot. But cheer up. The freezing process shouldn't hurt the Nano-agents. They'll be fine. On an atomic level, they are pretty hardy. They don't really need to be in a vacuum. The field is just a good way to contain them." If Stanford wants to pick up the project in 6 months or a year, I think I would be willing to start in again. And if I just happen to stumble upon pictures of the memory, I/O and CPU agents, then I'll posit a hypothesis and support it one way or the other with the factory functioning". She laid it all out nicely for them.

"Devious. I still say the truth would make a hell of a Nobel speech. And I want top billing", Joe told her as they drifted off for bed. They left everything hooked up and a mess. Plenty of time in the morning, right?

Chapter 5

May 25 0551 2010 Monday, Stanford, Ca.

Angela drove them over in the Lexus. Naqeeb quiet in the passenger seat. Joe grousing in the back with the box. "Who knew there was a 5:00 am as well?" He asked to no one.

Harold and Mike Johnson were waiting at the HAZMAT dock when they arrived. No one spoke as they handed over the box. Mike immediately took it into the HAZMAT area.

"I've got a long two days to get this broken down he said into the overcast "June Gloom" morning. The bay area was famous for its marine layer and foggy conditions. The clouds kept the daytime temps down around the bay. Inland, where the fog didn't penetrate, could be markedly warmer.

"Microclimates, brother" Naqeeb told Joe about the weather soon after they met. "That's why it's 62 in San Francisco and 101 in Livermore."

"Thanks, Harold, Angela told him. 'We need to get the laptop back into the project boxes today. Let me know if you get any heat." They said goodbyes.

Joe looked a little confused.

"Were you hoping for a shootout with the F.B. I? Naqeeb asked him.

"Kinda". Cops can't catch a clue".

"I might as well put in some work at the office, he told the pair. I have to deliver my baby to Thorson's office today anyway." Joe nodded and walked around the building to the Computer lab side.

Angela and Naqeeb walked around to get to the Nano side hand in hand.

Joe watched them go and shuffled over to his office. His cubicle really.

The cubicle farm had been his main workspace for the last three years. His senior project and Thesis work had mostly been done right here. True to the creative similarity of programming and artistry, there were three other students working away in at the desks. Asian males all three.

Joe was used to being a minority white male in this rarified world. Born in Hemet, Ca. the small desert town on the wrong side of the San Jacinto Mountains. The rich white lush side was Palm Springs and its sister cities of Indian Wells and La Quinta. Golf courses and shopping dominated. Hemet was the dry Latino farming side. Auto body shops and gangs dominated. Joe's father Herb was a welder until he died in 1997. Cancer.

His mother Mary took a job as an insurance actuarial and raised Joe and his sister by herself. The Italian Catholic mother had insisted her son apply himself. Really try. And it worked. His dad's death left a hole in his life that he filled with books and school.

Natural intelligence combined with an inner drive made him high school valedictorian. He hit UC Riverside with a partial scholarship and a pet snake named Estaban. Completing the undergraduate degree in two years he applied for an in-state program at Stanford still with the snake. It was perhaps his proudest moment, telling his mom he was accepted for a full scholarship. She beamed. "Remember where you came from and where you are going was all she said. "And you are not leaving that snake with me!"

Mary would be here the 27th. They had a family dinner scheduled on Thursday then the Friday afternoon conferment ceremony. Since it was only 61 Engineers and Computer Science candidates and their families, the ceremony was being held in the main church on campus. Mary had experienced with the whole walk the labyrinth thing and sit in the stadium for 4 hours circus from two years ago.

This was more personal. Joe snorted at his computer terminal. He kind of wished she was coming Wednesday night to the Kitchen for Naqeebs party. Joe thought Mary could referee what was shaping up to be a bloody battle. He was looking forward to the fireworks.

His phone vibrated as he logged onto his school account. He did not recognize the 202 area code. Somewhere back east?

"Hello."

"Joseph Smithson?" A male voice asked on the line.

"This is he", he said in his best adult voice.

"Did I wake you? The voice asked. Sorry, I know you guys keep odd hours and I'm back east and I took a chance." Running on.

"No worries, Joe lied. I'm an early riser- Mr...?" It hung.

"Jones, Robert Jones. That's great, he said. "Listen, Mr. Smithson, I work for New Star Technologies. We are a beltway IT company and we happen to have one of our recruiters in the area and we wanted to discuss your future plans?"

Jovial, collegial, reasonable. Joe didn't trust him at all.

"Boy, you guys move quickly. Stanford hasn't even published the list. Yet."

Jones laughed. "Things move quickly in the business world. We always have to be eyeing new talent, and right now we have our eye on you!"

"And I like to be looked at"! Joe enthused. He typed madly at the computer.

"Our rep, Stacy Brown, female by the way, will be e-mailing you to setup a meeting if you have some time available?" Jones let it hang out.

"Well, she can buy me lunch today or even breakfast. I happen to be open."

"Fantastic!" Jones rapped back. "I'll have her set it up then right?"

"Super, Bob!" Joe laid it on thick.

"Excellent. I'm looking forward to Stacy's report and I hope we can come to an arrangement." They said their goodbyes.

Joe repressed a gag reflex. Okay, let's see. He popped a CD into the computer drive and ran a program. He minimized the main screen when it came up. Joe typed away as he ran a search program on New Star. Calling up the company website he ran a second program in the background while he looked at the portal. Standard IT services company. He watched the screen as his e-mail blipped. There it was.

He dropped the e-mail into the first program he had open and clicked on the email icon. The innocuous page had an attachment that contained an invitation to a 9:30 breakfast meeting at the Westin Palo Alto. 2 hours from now. The attachment also contained a little malware program. He studied it.

Looking glass program. It allowed a remote user to see what he was doing on his computer. Next gen above a key logger.

He thought for a while and ran the second program. Joe e-mailed Stacy back accepting and saying he was super excited to meet her and explore new opportunities. He left both programs running and logged onto a third guest machine in his cubicle. He had a good idea what was going on and now he wanted to have a little fun. He cut and pasted into a file for half an hour or so. Not the most elegant hack ever but harmless.

He moved between the first two screens copying info into the third file. He was using technology at a furious rate he thought snidely.

He finally dropped the third file into the first program and ran it. Done. At 8:50 he logged off everything and grabbed two discs off the desk.

He went to Professor Thorson's office. Handing over the discs, Thorson reminded Joe of the release and non-compete clauses. Joe knew.

"Did Lines get to Rome Okay?"

"E-mailed me this morning. Said he took the earlier train."

"Smart man." Joe left.

Thorsons parting words rang in his ear: "Don't worry Joseph, They'll come around soon. You're too good for them not to."

Professor, they were sniffing around already, Joe thought. He hoofed it outside and caught the campus bus. It was crowded when he got on, but by the time he exited 15 min later at the Arboretum Grove there was no one else riding. At Galvezrd he took stock.

Stanford Stadium was a low-slung red brick bulge south of him. The lush green grove to the north. East lay El Camino real and his destination. The busy road was actually the eastern boundary to the grove of trees. As he went into the park like setting he had the place to himself. Half way thru he cut right and followed the sound of cars. He emerged from the beauty of the trees into the sight of a California Pizza Kitchen and a bike path leading under the road. He went under. He came up into a crowd of people going to and fro on the Sutter campus of the Palo Alto Medical Center. Joe knew the area well.

His job at Peetes was only four blocks up on Homer across the Cal train tracks. He rode this area a ton on his bike. He crossed Wells ave. getting to the Westin at 9:15.

He wanted to blend in but he didn't. Jeans and a tee shirt were fine at school but they didn't fit with the business crowd. He looked over the lobby carefully.

He couldn't be sure but he had suspicions. Crossing the lobby he scanned the breakfast crowd. He needed to be right on this one. A deep breath as he spotted what he thought was her and went in.

Joe dropped into the seat across from the woman and said, "Hello Ms. Brown, how is the NSA treating you?"

The attractive black woman looked up from her phone and said: "Excuse me?"

Gulp.

"Okay of that's the way you want to play it, I'm leaving." He got up to go.

"Sit," she said quietly.

Joe sat. "Call em off," he told her.

Stacy tapped her phone. Joe didn't bother to look around. He concentrated on her.

"What happens to the guys who don't figure out it's a test," he asked.

"Some of them are our best people. They come in believing the cover story, then we hit them with it and they stay and fit in beautifully. Truthfully, every candidate who has figured out it's a test beforehand has been a giant pain in the ass."

Joe nodded. "That's an interesting word *truthfully*. You've lied to me, tested me and hacked my computer. Truthfully is not one of the words you get to use." Huffing he sat back.

"Fair enough, she answered. We should eat." Catching the waiter's eye, she ordered for both of them.

"One of my colleagues caught your dissertation. He was impressed."

"Dark suit in the back?" Joe asked. Nervous.

Silence as she measured him while he watched her. Lanky, white, brown hair handsome watched athletic, black haired and brown skinned beautiful back.

"If you put some work into the predictive matrix you might have something in four to five years."

"Stanford owns the software," he told her.

Stacy spread her hands. "Come on Joe. Half the people who work for us came from Stanford. And Berkeley. And MIT. We can get around your release."

The food came and they ate for a moment in silence while Joe considered.

"That's all I'd be doing? Working on my software?" asked hesitantly.

Smiling widely she said, "What do you want to be doing Mr. Bond?"

Joe had the grace to blush.

"It's the line from "Good Will Hunting". "The movie", he told her. "They approach him about a job and he says to his friend "Sure sit in some windowless office doing long division all day."

Nodding, she asked: "And you'd want something more challenging? With a little fresh air?"

"Maybe… just who is putting out the data to you on that phone?"

"A giant pain in the ass like yourself."

Joe pushed back from the table and stood. "Thanks."

1) Give me a couple weeks. I need to think.
2) Don't fuck with any offers coming my way. I'll know if you do. I don't expect a million but a Ph.D. from Stanford ought to be worth 2 or 3 legit offers.

"Fair enough. We'll be in touch."

Cool. Cool. Cool.

Joe left thru the lobby without looking back. He nodded to the one guy he had marked on the support team. *Thought he had marked.* Shit! What now? He wandered back to his office in a daze.

Angela and Naqeeb traversed the halls of the Allen building that morning in a similar daze. They hadn't gotten caught. In fact, just the opposite. A regular morning packing the lab was all boring stuff. She floated until the afternoon. Her heart leaped when professors Lemov and Tompkins dropped by her desk. Naqeeb chatting on the phone a few cubes over.

"Ms. Chin, Professor Lemov said, I was wondering if we might have a word?" Are you well my dear, you look pale?"

Angela gutted down the impulse to blurt out she was sorry and guilty and they never meant to do it and...

She nodded yes because she didn't trust her voice yet.

"We were wondering how your job prospects have panned out this last month?"

"Any hot prospects?" Tompkins put in.

Angela let out the breath. "Jobs? No, none. I haven't really been looking." "I'm not sure where I want to go yet..."

Tompkins seized the opening. "Why go anywhere at all?" You could stay right here in the family". We have some excellent chip design startups that could use your skills or a battery initiative that looks like it's going to bear fruit. The battery team is developing..."

Lemov sighed and butted in. "Angela you are welcome to join one of the start-ups or ideally join the grant team working on the Nanotech follow up. I think you could garner some serious research money and complete what you started. Your talent combined with our facilities and you could work on the project for five or six years. I see four or five papers you could publish right now." "I see that as win-win. Doing what you love in a place you love," He concluded.

Her head spun. Did she want to stay at Stanford? She didn't really know.

Angela stalled them. Looking at her two professors and colleagues, she said, "That's a pretty powerful offer. Can I have some time to think about it?"

Tompkins bounced up. "Sure, sure. Take some time. I think HR is going to send you some information on the openings and the application process okay?" She nodded them out.

As the two went to approach others in the group. Angela leaned over into Naqeebs cube.

He looked up at her. "What's up?"

She related the talk with Lemov and Tompkins.

"Uh, congratulations?!?" he said trying to read her. "Are you happy or sad?"

"Both. Neither." I'm not sure." Sitting on the only available chair in the cluttered office space she faced him. "On the one hand staying to do more research would be great. We could follow up on last night. He noted the "we". "Not much money though, she said not that that matters too much."

"Hold on a moment. Have you ever written a grant proposal? "

"No."

"You get to write yourself in a salary. Personnel costs are just one line item. X number of people for Y dollars for Z years." He paused. "Theoretically you could write yourself in for a million dollars a year for five or six years."

"Really?" She stared at him.

"I'm not saying you should do that, but paying yourself pretty well could be in the cards." A materials engineer in silicon valley starts around 80,000 per year and moves up to 150,000 with experience," he told her.

More contemplation. "Stanford has grant writers. Someone will get with you to define the statement of work, get personnel costs, lab facilities, and other requirements. Then write it all up." "You're probably looking at 120 per year" he finished.

Silence stretched as she rolled the info in her head.

"You're not down with this huh?"

"Am I hiding at Stanford?" she asked suddenly.

"Hiding? No. The University does a good job of wrapping up its stars in a comfortable environment and working them hard. "That's you by the way." Same as any outside company," Naqeeb said slowly. He wanted to support her on this. "Why don't you field other offers?"

"What other offers?" she said looking around.

"Honey, it's 2:00 pm on Monday before the ceremony. I'll bet you have 5 offers before this time next week."

Moving her gaze up to his face, she said: "what about you?" "Anything?"

"Nope not yet, he said confidently. I'm sure something will turn up".

"I'll write you into the grant," she started loyally. "They'll need software services…"

"No, he said cutting her off. Don't take it just to put me in there and keep us together." "Don't worry about me. I'm fine and can work where we decide to go. Bay area, DC, the research triangle, Boston. Anywhere"

She nodded.

"Let me hear you say it: "I will not stay at Stanford just to make Naqeeb happy"

She dutifully repeated it back. They hugged.

Joe found them like that when he arrived at the desk, breathing hard.

"Good you didn't get arrested!" We need to talk"

"Whoa. Slow your roll. What's going on?" Naqeeb asked.

Joe related his breakfast meeting and job offer from the NSA.

"Robert Jones, Stacy Brown. God, who makes up their fake names anyway a third grader?"

Joe scrambled to find a chair. He dragged it into the cubicle and sat.

"I finally thought it might be rude to separate us while they grabbed you guys," he told them.

"Wait, you had this meeting at breakfast and are just now checking on us?" Angela said mockingly.

Joe sheepishly looked down and mumbled: "I just now made that connection sitting at my desk. I came right over when I figured it out," he shouted defensively at the end.

"Relax, Jason Bourne. We're cool. Besides Angela has news too."

Joe looked at her.

"Stanford has offered me a job. Well, technically they want to explore possibilities." She said

Joe looked relieved again. "Okay, yeah, you guys are good. They wouldn't offer a job if they knew about this weekend. A job huh?" He considered. "Doing what"?

"Same old same old, Naqeeb jumped in. "So it looks like the NSA wants your snooping program huh?" "What was the agent like?"

"Scary. Dracula crossed with... Joe thought hard stumped for a second. Who's that actress, black lady? From Fantastic Four/ Silver Surfer?"

"Halle Berry?" answered Naqeeb.

"No. "Kelly Washington? Kerry? Kerry Washington". "Yeah, that's it. Dracula crossed with Kerry Washington." "Scary".

"What are you going to do Joseph? Angela asked seriously.

He faced his two friends. "I'm going to put them off a while and then take it I think." When he said it out loud it felt better.

"Joseph, the NSA is spying on Americans", she began.

Joe held up a hand cutting her off. "Google and Amazon know as much about you as the NSA. Who do you think is providing the data?" If I worked for one of them I would be bored silly trying to sell more products and ads. This way I get to fine tune my baby and use it for truth, justice, and the American way!" He leered.

"And you get to work with Halle Berry". Naqeeb put in.

"No, Kerry Washington. But you are missing the point."

"Point is you are thinking with your dick."

"I am not thinking with the little head, Joe scowled. It may not be ideal but it's the best job for me."

"Don't sell yourself short," Naqeeb told him.

They looked at each other. "Don't worry man, it's all good."
Cool, cool, cool.

Angela gaped at them. She marveled that the two best friends could accept monumental changes in each other lives that casually. That easily. She had no doubt that Naqeeb and Joe were ready for the inevitable break up of their little triad. Just like that. How could that possibly be? They needed to talk thru their feelings and reassure each other that the bonds they had established would be okay.

Joe stood. "Hey, let's cut out." I'm finished and I want to play some Xbox before Renee comes over.

Naqeeb readily agreed. Men! Angela tried to hide her exasperation. Two years from now they would see each other and pick up right from here like they never parted. Unfathomable! What's that word from that stupid movie they kept quoting back and forth? Inconceivable!

They packed up and headed home. Settling into the little place, Angela started reading while Naqeeb hooked up the Xbox and the computer.

"Why did you do that, Joe complained. That's why Halo is glitching"

"My graphics card is way better than that old Xbox console," he started.

"You think you're so cool hacking the console and re-routing the output," Joe said.

"Maybe I'll get a job at the NSA too. They need someone who can program."

"Yeah, we need waiters at the cafeteria."

"Fuck you racist."

Joe smiled.

The TV was showing a local news program. Margaret Pullman was being interviewed. Angela glanced at the flatscreen. "What's Pullman doing?"

"FastFac is putting out a new device", Naqeeb told her.

"Wow that's a tough multi-task: CEO of a tech company and run for Senate," Joe said.

They listened while Pullman waxed on about the Scimitar. File footage showed the President, head down using the device in the "Text Walking Posture" as he surfed his messages.

"Dude, it is so cool to see the President *Text Lemming*, Joe crowed.

Angela smiled, "no he is a *text zombie*".

"Are you going to vote for her?" Naqeeb asked her seriously.

"Why would I vote for her?"

"Well, she is part of the sisterhood." put in Joe.

"But she is a Republican. I lean more Democratic. I'm supposed to vote for her because we share a vagina?"

"Whoa!" "You can't say the V word!" "That's kind of sexist and it offends me to hear that kind of objectification of women!" "Aren't we more than our parts!" Both men laid it on thick. Talking over each other. Then they spoiled it by giggling.

Joe ended by reminding her that she didn't "lean" democratic, she was a full on liberal.

"And you're not?"

"I voted Republican in 2006, he reminded her.

"Me too", from Naqeeb.

Angela sat up really ready to lecture them. "You two idiots voted for Arnold Schwarzenegger and look what it got us? A serial grouper in Sacramento!" They immediately started doing Arnold impressions: "We must improve the lives of every Ca Li for Ni Ian." "It's not a tumor!" "Hasta La vista, baby"! "I'll be back!" More giggling from the couch.

"Well I'm not voting for her, Angela concluded. She'll never beat Boxer, not in California."

"You know Pullman is speaking at the ceremony Friday", she said. That shut them up.

"I thought it was Pelosi? " Joe said.

"I guess she is sick," Angela said tartly.

"Sick of everyone asking her if she tipped off her husband on those stock deals," Snark from Naqeeb.

"That's horrible to say about her!" Angela reproached him.

"Kidding dear!"

Angela watched the rest of the interview in silence. Pullman was smiling widely at the end. Joe and Naqeeb started chanting from the couch: Halo! Halo! Halo!

She acquiesced and the TV was shifted. 2 hours later they started chanting Pizza! Pizza! Pizza! Honestly, it was like living with five-year-olds.

There was one advantage to the Halo explosions and carnage she thought as she dialed the phone. It kept them from dwelling on the potential explosions and carnage of Wednesday night. The families were getting together for a party. Her parents. Naqeeb's parents. Together in one room. Wednesday night should have its own ominous chord of music like a horror movie. Duhh duhh duhh duhhhhh!

CHAPTER 6

May 27 18:51 2010 Wednesday, Fremont, Ca.

The horror movie thought kept intruding into her brain. Wednesday night loomed and lurked and finally rolled around. Angela said as much in the car with Joe, Kelly, and Naqeeb. Saw, Paranormal Activity, The Fourth Kind, The Grudge, The Haunting, Evil Dead. Joe and Naqeeb flinging out horror movie titles like a game of Scattergories.

Kelly smiled. "Shut up you two." She could see Angela was nervous.

Naqeeb drove up the 880 to Fremont. Rush hour traffic was still clogging the freeway. He concentrated on the road and tried not to dread the coming meeting. The family party was his parent's idea. Naqeeb had a number of Aunts, Uncles, and Cousins that wanted to congratulate him. His mom and dad also wanted to show him off like a prized pony. He agreed only if he could invite Angela and her family. And Joe.

You'd have thought he was suggesting a satanic ritual instead of inviting his girlfriend's family to the party, the way they carried on. *Drama from his momma* was how Joe put it.

Shireen and Khushal swung them around. Naqeeb had quickly applied the pressure to his siblings. He pointed out that if they didn't support him in this they were out of luck on their own battles. The Children's Unified Front prevailed.

Exiting on Mowry, he mentally prepared for battle. He was dressed in a traditional Sherwani and Churidars. The cream colored Sherwani was a long jacket with gold embroidery. The stiff Nehru collar was giving him fits. The Churidar pants were tight and dark brown. A matching silk shirt completed his ensemble. Kelly whistled appreciatively when he came from the back bedroom earlier. Naqeeb sighed. "You guys clean up nice," he complimented. Kelly was in a print dress that showed her figure, while Joe was in his one and only dark suit.

All three watched as Angela emerged from the bedroom. The high-necked, form-fitting red silk dress was only half the game she was bringing tonight. Gold and lots of it adorned her neck and arms. Her hair was swept up and held with a gold and jade ornament. She nodded at Kelly who smiled at her. She looked gorgeous.

"Shit, we are never making it out of Fremont alive," Joe muttered. They were off.

Now as Naqeeb slowly cruised up Mowry, Joe mistook his silence for reminiscing.

"Glad to be back on the mean streets of Fremont?" he kidded his friend. "Take a good look ladies, every single significant event in Naqeebs life took place within a twelve block radius."

Brier Elementary, Walter jr. High and Washington High school, right Naqeeb?"

"We used to live in that subdivision on the left," he confirmed as they passed Logan st.

"He got his face slapped for the first time at Lake Elizabeth right up this road," Joe crowed.

"I thought your parents lived in Niles?" Angela saved him.

"Yeah, now, he told her. "I never lived in the Niles house. We grew up back there." He pointed back over his left shoulder. "I was a sophomore in college when they moved into the new house."

A quick right on Fremont blvd. and another into the parking lot brought them into the Fremont Plaza Shopping Center. The Indian Kitchen restaurant sat alone in the corner of the mall. It competed with all the standard American chain places. Khushal met them at the door as they went in. Dressed like Naqeeb, he stared openly at Angela and Kelly.

"Dude!" Naqeeb snapped him back to life.

"Yeah- they're all in the back room. All of them," he emphasized. Angela laid a hand on his arm. "Khush, you remember my sister Suelyn from the party at our apartment right?"

"Sure, sure".

"She's bringing my parent's tonight. Could you clue me in when they arrive?" Khushal nodded agreement and stood guard duty proudly.

"Nothing to do but do it, Joe said taking Kelly's arm. Naqeeb and Angela held hands as they went back.

The Indian Kitchen was quite large. The front side held 30 or so tables while the banquet room held 300 people comfortably. The dinner rush was well under way. Waiters and busboys nodded to Naqeeb as they moved into the banquet room doorway.

The bright "CONGRATULATIONS" banner stretched across the ceiling. 50 assorted family members broke into applause as the group entered. Naqeeb and Angela were immediately swallowed up by the younger cousins, jostling to say hello. Joe and Kelly slid off to the left, spying a friend. Shireen came up the pair and smiled hello. She was dressed in a blue Sari with more embroidery. Stunning silk. She gave Joe a quick kiss hello and hugged Kelly. "I have a table for us right over here" she pointed and led them over.

"What's the setup?" Joe said over the music.

"Food is against the back wall, buffet style. Any waiter can get you a drink. Dancing in the middle later on." Shireen motioned for a waiter to help them. They ordered waters. "The real action is going to go on right there", she pointed.

Two tables were placed in the center with a clear area around them. One contained a number of Naqeeb's Aunts. The five older women were seated wearing richly detailed saris and similar facial expressions. Frowns. There was one open seat at that table.

The second table held Mr. and Mrs. Al-Meri and three open spots. That didn't look good.

"Joe, don't you want a real drink," Shireen asked as the waiter brought their glasses.

"Nope." "I'm pretty sure I'm going to have to throw myself on a grenade tonight so I want my wits about me."

Kelly laughed and said, "Come on! It won't be that bad! And don't you do anything stupid." She turned to Shireen. "Have your parents met Angela before?

The girl shrugged. "Yes, once. It went okay. They thought it was a normal girlfriend thing that wouldn't last. "Then you moved in together", she accused.

"Not me sister", Joe protested. "Come on Kel, let's get some food and say hello." He grabbed her hand and led her over to Naqeeb's parents. He put on his best smile. "Mr. and Mrs. Al-Meri thank you for inviting us to the celebration tonight. "This is my sister Kelly. Kelly these are Naqeebs parents."

The Al-Meri's shook hands politely with Kelly while congratulating Joe on his dissertation acceptance. Mr. Al-Meri had one eye on Joe, one eye on Naqeeb and one eye on the doorway. He asked Joe about job prospects. Joe put him off with the "irons in the fire speech. Mrs. Al-Meri nodded gravely, dismissing them.

"This could be bad," Joe told Kelly in the buffet line.

When they returned to the table, Naqeeb was sitting with his Aunties while Angela sat with the Al-Meri's. Everyone looked uncomfortable.

Angela was jumping out of her skin. She spotted Khushal in the doorway nodding at her. Here we go she thought. She rose and told Naqeeb's parents that she thought her parents had arrived and she would escort them to the table. She stumbled to the doorway as the group arrived. Suelyn gave her dress and jewelry a look as they hugged. She was in a basic black cocktail dress. And a huge gold and jade necklace. A girl had to show some bling. Angela's parents Mei Tam and Dr. Ri Chin were dressed conservatively. The only nod to their wealth was her mother's ring. The large diamond sat prominently on her thin graceful finger.

Angela embraced them whispering a "be nice" warning. She guided them over to the waiting Al-Meri's. She watched Suelyn maneuver out of harm's way. Coward. Her sister joined Joe, Kelly, Shireen and Khushal to watch the byplay.

Angela presented her parents saying, "Mother, Father, this is Naqeeb's parents, Sanjay and Aludra Al-Meri. Mr. And Mrs. Al-Meri, may I present my parents, Mei Tam and Ri Chin." Stiff hellos and handshakes all around. Angela sat looking for Naqeeb and help. No luck from that quarter, he was dutifully playing waiter getting his Aunties food.

Sanjay gestured and a waiter appeared bearing a tray. "A drink to celebrate our good fortune". Four small glasses of Tadi were placed on the table. None was offered to Angela. The parents toasted the children's success with palm wine.

Dr. Ri nodded appreciatively. "That's very good."

"Thank you", from Sanjay.

"Thank you for having us over to your fine restaurant, the doctor added back. "It looks very busy tonight".

Sanjay waved away the compliment. "We have been very lucky. We started when the area was just taking off with the NUMII plant."

Angela knew NUMII made auto engines for a car manufacturer. She supposed the United Auto Workers ate large lunches filled with Tikka Masala and naan.

"Yet you were able to make it thru the NUMII shut down and hold on until Tesla took over though" Dr. Ri queried.

"Aludra always plans for emergencies," Sanjay said. His words caused Angela to focus on the two women. They were eyeing each other silently. Sizing each other up. Uh oh.

"Sanjay, Aludra said quietly, "perhaps Dr. Ri would like a tour of the kitchen."

On top of that Mei Tam told Angela, "Daughter, go help Naqeeb." Angela cringed at how quickly the "yes mother" was out of her mouth.

The men and Angela left the table. The two women began talking.

Joe and the others watched in fascination as the play reeled out.

"It looks like the Korean Nuclear talks", Joe said to nobody.

Suelyn disagreed. "This is much more complicated."

"What's complicated", Khush said, "they love each other."

Shireen and Suelyn shook their heads. Shireen told her brother, "There are lots of areas to work out."

"Such as?" Joe prompted.

In a traditional Indian arranged marriage the eldest sons' wife is expected to take care of her mother in law. "Cook for her, care for her, defer to her," Shireen said.

Kelly blanched.

"It's much the same for Chinese", Suelyn informed them. "Mostly I think they are trying to figure out the children."

Joe gaped while Shireen agreed. "Which religion, what customs, where they will live. Lots to work out."

"Khush, you getting all this?" Joe asked him. Khushal looked ashen.

Joe watched as the talking continued and plate after plate was delivered to the table. Some sampled, some just admired.

Kelly leaned into him and said softly, "think mom could take them?"

"Whew. That'd be a battle royal", Joe said. Mom could hold her own." "You know he said addressing the rest of the group not just Kelly, "If you think about it the two women are really similar." Smart, determined, caring. And used to getting their own way on things".

The women started debating their own mother's strengths and weaknesses.

"Stubborn both of them, Suelyn concluded. "But hey, as long as they focus on Angela I am off the hook for grandkids." There was a good reason she lived in Sacramento.

Ri and Sanjay rejoined the women. Naqeeb and Angela were summoned from the Aunties table to join the parents. All six sat stiffly talking. Around them, Naqeeb's cousins cranked up the party. Music got louder and dancing started. Shireen was summoned to the head table. She came back eyes downcast. "They want to talk to you," she told Joe.

"Why me?"

"They saw us talking together and they think something's up."

He stared at her. "Been using the white guy as a way to tweak your parents?"

She protested. Kelly told him to go and be polite.

"Yeah, yeah, let's go." He went to talk to the group of elders. He endured the grilling. *No, he had no designs on Shireen. Yes, she was a lovely girl who anyone would be lucky to have.* Angela's parents reminded him that Suelyn was married.

Joe looked at his friends. They appealed to him with stares. "I'm just getting free medical advice. I have a bum knee he told them." No one seemed to care about his academic credentials or his job prospects. On the way back to his table he decided that he had been way too passive at this party. Time to enjoy himself. He started with a dance. With Shireen for good measure. Then a beer. Then a dance with Suelyn. Then Kelly. Then more beer. Ninety minutes later he was sweating and a little buzzed. The Aunties table glowered at him. However, the head table seemed a little thawed. Naqeeb was talking more. Angela even had a brief smile. Maybe the Six-Party Talks had produced a Nuclear Accord.

Kelly swirled back to the table. She had been dancing with Naqeeb's cousin. His taller more handsome cousin. "I can't wait to see mom's face when you bring him home," Joe teased.

"Relax, Sumeet just likes to dance. And that was a little racist." Ya huh.

Joe looked at his watch. 9:30. He owed Renee a call. Shireen went by and he grabbed her. "Hey, come her a second." Shireen glanced at the Aunty table before joining Joe. "Whats going on there," he gestured to the head table.

"Ahh. The parents finally found an area of common ground: you.

"Me?" he said.

"You. You need to settle down. You need to apply yourself. You need to be more serious. You need to stop drinking."

"Three beers at a party and I have a drinking problem?"

"I think it's four, Joe. The Aunties don't miss a trick." Shireen hurried on.

Good. Joe caught Naqeeb's eye and gestured to the door. His friend joined him outside. The night air was cool after the body heat of the party.

"Howzit?" Joe asked.

"Thanks, dude" was all Naqeeb said.

"Figured you'd need some cover." How are Angela's parents?"

"Tough, Naqeeb muttered. "Well her mom is. Stubborn."

"Yeah, that's true, Joe agreed. Not unlike someone else we know hum?" Naqeeb silently agreed. His mom was the same.

"You can't wear away the rock with a single raindrop," Joe said.

"Confucius?" he asked.

"My mom actually," Joe laughed.

They returned to the party to find the Aunties all standing exchanging goodbyes with the Chin's. Angela looked exhausted and relieved and angry.

Detent reigned while the simple act of leaving a party stretched to 30 min. Angela escorted her parents and Suelyn to their car and returned with a stricken look on her face. Naqeeb felt for her. The couples made their own escape a few minutes later, as the party wound down. Naqeeb slowly drove the four out of the parking lot.

"Holy shit we lived," Joe exclaimed.

"I'm glad someone enjoyed themselves," Angela muttered. Naqeeb squeezed her hand and said nothing. Kelly spoke into the silence. "Did you guys get anything to eat?" "About two bites," Naqeeb answered.

Kelly held up a bag. Shireen sent this along". He smiled at her in the rear view mirror. Joe butted in. "What did your parents say, Angela?"

"Not too much. My dad was impressed with the restaurant and he likes Naqeeb and his parents. My mom..." she let it hang.

"They practically suggested splitting up the grandkids. They want two each." Muttering his outrage Naqeeb merged onto 880 southbound.

"Aren't they getting ahead of themselves, Kelly said. "You guys aren't even married yet, you don't have kids. You don't even have jobs yet really." She added.

No response from anyone.

Joe waited to see if there was any more talk. He decided to bring up the secret.

"Did you hit them with the big news?"

"No, we decided to wait... "Angela started.

"What news?" Kelly asked over her.

"I got an interview with Google tomorrow, "Naqeeb told her.

"That's fantastic!"

"It's only a matter of time before Big Blue hits you up, Joe told him. "I don't know why you decided to wait to tell them. It would ease some of the pressure."

"The problem is it's our life. Not theirs." Naqeeb said stubbornly. "We get to decide what goes on". Besides, IBM has Watson up and running. They're pretty far along the A.I. path."

"That's why you have Google over a barrel." Get stock options as a signing bonus," Kelly advised with the air of a seasoned negotiator.

Angela smiled briefly. The party was nerve wracking. Her stomach was still in knots. She only got 30 seconds of conversation with Suelyn. She had been trapped between the two moms. The Tiger of her mother versus the Elephant of Aludra. That was unkind, she thought. Aludra was really a pretty woman. Just powerful. And determined. And she never forgot anything. Imagine negotiating the raising of grandkids? Kelly was right, we aren't even married yet.

She thought they would be someday. But not right now. It wasn't that she didn't love Naqeeb. She did. It's just that they wanted some time and space to live their own lives before jobs and marriage and kids complicated things. And things were going to get complicated. She cringed thinking about just the religious aspects. She was a practicing Catholic. She wasn't particularly devout and neither were her parents, but her faith was important to her. Naqeeb was similar in his feelings as a Muslim. The real sticking point was a duty. Angela's duty to her family and Naqeeb's to his. And transversely to each other's families after they were married. She didn't mind being nice to her mother in law, but no way was she going to defer to her. Not about kids and not about Naqeeb. She clenched her jaw at the thought.

Naqeeb squeezed her knee. He always knew when she was upset. He just sensed it and comforted her. She loved him for that. She looked at him bleakly.

"Relax, he said. "They don't own us. We control our lives, not them." "I know it's harsh but if they want to be part of our lives they have to play by our rules. "Ditto with any kids we may have".

Joe cackled from the back seat. "That plan will never work!" "Did you see your mom?" She'll chew you up and spit out your bones, I'm telling you…"

Kelly cut Mr. Sensitivity off. "Did you tell our mom about Oracle yet?"

Joe stopped and the smile fell off his face. "I was saving it for dinner tomorrow."

"Yeah, tough guy, Naqeeb needled back. "Try telling her you might take a job in Seattle. Let's see how you stand up to parental pressure. Go ahead."

Kelly tried to make peace by sidetracking the conversation. "It looked like your dads got along," she told Angela.

"They did I think. They don't have much in common but they are like all men. They have it easy. Go to work and then come home and be pampered."

That started them whole male/female debate. It lasted until the apartment parking lot. Kelly got in the last shot. "Babies!" "If men had to give birth the human race would end!"

The tired couple said goodnight to Kelly and went up. Joe hugged his sister and dispensed brotherly wisdom. "There's no toll eastbound on the San Mateo Bridge. The 92 goes right to campus, so head straight home."

"Yes, Joe" she replied suffering the obvious advice. "See you manana with mom in tow, Okay?"

"Night" She drove off.

The lovebirds had gone right to bed when he got into the apartment. Naqeeb had his interview at 9:00 am. Joe thought it would be a six or seven-hour grilling. Two or three different panels of people asking various levels of technical questions followed by lunch. Then the tour of the fabulous facilities (Look, a rock wall!). Then a smooth talker from Schmidt's office comes in for the hard sell. They might even trot out Page or Sergie Brin to shake hands to seal the deal. If Oracle played him like that he was demanding a ride on Ellison's yacht. Damn straight.

He flipped on the Xbox. It still ran thru Naqeebs computer. He didn't have his interview until Monday the 31st. Killing people in Halo was an excellent way to deal with the stress of job hunting and the party. And he was mowing them down today! He hated to admit it but Naqeeb was right about his graphics card. Crisp and clear and no glitches. Cool Cool Cool.

He joined an on-line mission and was messaging back and forth with his group. 2 am saw him bleary eyed, messaging with some foreign dude who did not speak English well. Enough! Bed time!

He logged off and unplugged the Xbox from the modem. He stumbled into bed. If he fell asleep on the couch the cat would inevitably curl up on top of him and Naqeeb would find him in the morning cuddling with catpuccino. His street cred couldn't take the hit.

Thursday day passed in fits and starts. Joe rolled out of bed at 10 after turning down a shift at work. He had some time off and he was going to enjoy it. He missed Naqeeb leaving. Angela was on the phone talking seriously with someone. He shot her a raised eyebrow question and she mouthed back: BERKELEY.

Yeah, no shit Berkeley, Joe thought. They liked Nanofabrication as well. And they had grant writers too. Angela was giving them hell. "Was the Grant money in hand?" "How much for how long?" What was the research team makeup?" Lab facilities?" "Were they looking for a software person?" "Would she have autonomy?"

Joe smiled. Stanford had created a monster. She chewed a fingernail after hanging up.

"Well?" he said.

"They mentioned an "Honorarium" in addition to the grant salary."

"Jackpot!" He crowed. He flipped on Sport Center. Giants fell in the nation's capitol 7-3. Barely above .500 at 23-22. Crap, not again. Bouchy better gets the boys motivated. "Will HALO bother you?" he asked.

"Not any more than normal," she replied. "Go on, have fun."

"Thanks, mom!" He flipped over to Xbox while Angela read and made some lunch. Joe scrolled thru the messages. Man, who was he talking to last night? Some of the messages were words and numbers. Weird shit. He played for another two hours. The gameplay was good but the message system kept screwing up. Joe couldn't figure that. He finally tired of dicking with it and tried to fix it.

He opened Naqeebs laptop to look at system specs and the graphics card. What was the MPS for the data rate? MPS was Megabits per second. He also wanted to see the CPU usage and the clock rate. If they had a synch problem that might account for the messages getting garbled. But wouldn't the graphics be all scrambled as well? He checked the TV refresh for good measure. All three were at 30 Hz. Okay. Wait what?

That was way wrong. The TV should be at 120 Hz. The CPU at 360 GHz. While the graphics card must be 240 GHz at least. He switched the TV back to Sports Center. Normal picture but still reading 30 Hz. He tried changing screen aspects. No luck. The picture looked perfect... He tried the oldest computer trick in the book. He toggled the on/off switch. All three got the Edison reset. They came up just fine. Computer, Xbox, and TV all working okay except for the messaging system. That was pretty much gibberish. He disconnected the Xbox from cable. He'd work it tomorrow. Angela asked what was wrong. "Just some weird shit on the Xbox."

"You mean besides the game?"

"Ha, Ha" "I'm hitting the shower, he told her. "More family stuff for me tonight." "Where's your mom staying?" she asked.

"The Sheraton on El Camino" he replied.

"That's nice!"

"Yeah"

"Where are you eating?" she pressed.

"We are hitting the Daily Grill in Union Square."

"Nice times two!" Angela said again.

"Yeah, I like the place. A little dinner then some window shopping, then BART back to Daley City." He concluded. "What are you guys up to?" he asked.

"Not much. R&R right here." She said.

"Good. "I hope he gets home before I have to pick her up. I was hoping to hear how it went. Joe looked at the front door.

"What time does she land?" Angela asked trying to figure for him.

Joe tapped his phone. "Southwest flight 4277 Ontario to Menneta San Jose- 3:52 on time."

She took that in. "That's going to be close to when he should get here. What did your software say about that flight?"

"It said to take that flight," he answered promptly and proudly.

"Really?"

"Yep. This is a good real world test case." Look at all the variables: 4 So Cal airports she can go out of. Three bay area airports she can come into. Highways, hotels, activities. Little Mary Smithson has to get from Hemet to here and back. Drive then fly then stay then activity then return then fly then drive." He listed the items she needed to accomplish. "What's the best way for her to do all that?"

"Best?" Angela asked.

Joe sat with her at the kitchen table and lectured. "Yes best. Best as defined by parameters. The best highway that is the shortest route with the lightest traffic around her travel times. "What airline and airports dovetail into that highway with the best on-time departure and arrival with the shortest flight time for a reasonable amount of money?" "What hotel provides the best value given availability and price versus a minimum level of comfort?"

Angela frowned at the complexity. "We just fly out of SFO".

"Of course," Joe said. "And that makes sense if you were heading to London. Direct flight for pretty cheap. No sense in making 3 stops to save 100 dollars right? But what if you could make one stop and then save 500 dollars?" She considered that. Now consider the weather in October on the peninsula."

She struggled and then said "Foggy".

He smiled. "Right. So if you had to go to London in October then it might make sense to go Oakland to Los Angeles then direct London. You save 500 bucks and you get there when you are supposed to. All you give up is a BART ride over to the east bay." My point is that all of that is predictable if you know how to work the variables."

Angela mulled that.

Joe said softly, "I can even have it solve for your Stanford versus Berkeley debate.

She scoffed.

"Seriously". "The data is out there." Both schools have a track record of getting government funding. Grant money all has to be reported. We would need to define the variables and the outcomes to find what works best tailored to you."

She thought about that. 'How would you go about it?"

Joe paused thinking, "Well, if the best outcome for you is to have Naqeeb working with you, then the variable which shows the University with the best record of allowing the project lead to define team members is the obvious choice." I guess there is a small chance both could be equal in that regard." Then it comes down to location, facilities, pay and other variables versus the one outcome that is equal." Plus you need to remember that they could both be equally good at allowing you to set team members or equally bad."

"So if neither one allows me to write him in then I need to decide if that outweighs continuing with the evaluation of both facilities", Angela said thinking out loud.

"Exactly." "It's all probabilities and weighting of variables." The key is information on you and the schools and what defines the parameters." Joe went on.

"That's fascinating."

"You know even if you don't want me to use the software, the definitions and personality sheets could help you solve it for yourself."

"Like a pro and con list?" Angela asked.

Joe shrugged. "That works for some people. My profiles are a little more sophisticated but that's essentially it. I just back that up with data mining to see if your wish list/profile sheet is honest." For a pro or con list to work you have, to be honest." For example: if you really, really, detest driving then any job with a long commute is going to make you miserable." "That might be a deal breaker." "But for most people, they can put up with some negative aspect because the job provides other aspects that they love: Co-workers or the type of work or the pay." With enough information about you, I can predict what you will do for a given problem." Joe concluded modestly.

Angela turned morose. "Where to work is ancillary with the problems concerning our families," she told him.

Joe put a hand on her shoulder. "Same approach just different variables and parameters. And more of them. Plus they are a little harder to quantify but doable."

Angela was skeptical "you could never quantify my relationship to my parents."

A scoff this time from Joe. "Sure we could. It's quantifiable in a million ways by your previous choices and history." He got up and walked around the kitchen talking animatedly. "Why did you go to Stanford and not Harvard?" I know you got accepted to both places right?"

She nodded.

"Why do you live here and not at home?" "Both decisions speak directly to your relationship with your parents." "And I can pull data from the internet to support or disprove my analysis and hypothesis of which direction you would likely go."

"Such as?" she prompted.

"Your Harvard and Stanford applications are both on file. Likewise the acceptance from both institutions. Your registration at Stanford is also on file. That shows your choice to stay in California near your parents. At 18 you lived at home and went to school. At 21 you went to the same school but lived here. You are on the lease", he reminded her. All that's available." All those choices speak to your relationship with your parents following a normal, predictable pattern of leaving the nest." See?"

"Some of it is supposed to be private!"

Joe shook his head slowly. "That is an illusion and you know it." Let's say hypothetically, that I can track you on the GPS function on your phone."

Angela looked alarmingly at him.

"Hypothetically!!" "But say that a pattern emerges that you spend three of four Saturdays at a Hillsborough coffee shop." "Working? "No." Meeting someone? Perhaps? Book club? No, Boyfriend? Perhaps, but there would be other activities, drinks dancing movies dinners." So who are you meeting? Daddy for a weekly coffee check in?" To back that up I need to link his activities to yours. But once I do, I can match the data and then up my correlation of the probability that your parents mean a great deal to you." "That becomes a discrete bit in my predictive probability matrix." "One bit mixed with thousands of others given the info I have and can get on you and I can predict what you are going to do." "It's all data mapping and linking."

Angela took all of that in and said: "Joe that's a little scary."

"I know what this is. I'm aware of what this can do. My main thought is that large companies are doing this already. Your search for cruise vacations on the internet and soon your spam file on your e-mail fills with cruise offers. Coincidence? Of course not." But there isn't anything really malicious in that. Kind of a: "hey, if you're interested, here's a deal!" And it can be used for good as well!" People and systems are predictable." That can allow us to anticipate problems and work around them." Angela was not going to stop him in his rant.

"Look at Carmeggedon."

"What?" Angela couldn't follow the non-sequitor.

"Remember in LA on Memorial day last year? They were going to shut down the 210 freeway to do some road work?"

She remembered.

"The press gave those dire predictions of gridlock in LA because everyone uses the 210 and the 405 and the 5 are normally nightmares." "That never happened and traffic throughout LA was way lighter than usual." "Why no carmeggedon?" Because people had the info and were able to adjust their normal patterns to work around the problem." "I can show you your little personal carmeggedons and help you work around them."

Angela chewed this. "Like Professor Line's train trip or your mom's flight?"

"Yep"

"Your call, he told her. I'll design some questions and forms for you and Naqeeb and run it thru to see what it says. You'll have your answers," he concluded.

Joe could see her waffling. She was under enough pressure. He eased her off the horns. "Hey- no worries. There's always this: My mom gives the best advice. She always says "don't overthink everything"!

"You love the guy right?

She nodded tearing up.

"Be together," he told her. "Figure the rest out as it comes." Joe gave her a brief hug and went to get ready to pick up his mom.

Naqeeb straggled in as Joe was leaving. "I'm wiped," he told them.

Joe angled by him in the doorway. "I want to hear all about it later."

"We'll talk tomorrow morning" Naqeeb confirmed.

Joe stopped and held a hand out preventing the door from closing. "Hey- Unhook the Xbox from your card. The stupid hack is ruining the game."

"Wha?"

"The gameplay works but the message system is fubared. Bye"

The couple sat on the couch. Angela let him rest his head back for a minute. He looked nice in his interview clothes she thought. "How was it?"

"Exhausting, he said. She let him tell it his own way. "Man they have a nice setup, he told her. That headquarters looks like a playground."

"Did you meet Page or Brin? She asked excitedly.

"Nah, they were in New York." I spent the morning with some HR teams and then the head of the AI project and I had lunch." Mark Zhou. He went to UT in Austin. Freaked me out a little to hear that southern accent coming from that Asian face."

He laughed a little at that. He looked at her seriously. "He's only 30. I met the rest of his people. We talked AI and programming and baseball for two hours." I might be the dumb guy in that group" he told her.

She hugged him fiercely. "You are not, don't be silly".

"We talked money," he said quietly. "An offer letter is supposed to be out on the first."

"Really? That's great!"

"You want to know?"

"Only if you want to tell me."

He smiled. "95K to start. A forty thousand dollar signing bonus. 20 in cash and 20 in stock. Options available at the end of the year based on milestones. Medical, dental, 401K matching. The whole works," he rattled off the list. Angela gasped. Naqeeb shrugged. "I wish it was my shrewd negotiating skills. Mark said he knew IBM would be calling and he wanted an answer before we met."

She looked at him. "You have to take that right?"

Shrugging again. "I wouldn't be working with you. And what if you end up at Cal?" "Then there is the whole deal with our parents." He broke off.

Angela rubbed his neck. "Hey- Don't overthink this. We will be together no matter what. I will support you in whatever you want to do. I'm proud of you."

He slowly lifted his head and looked at her brown eyes. "That worked". He got up and scooped her up in his arms. She laughed and nuzzled his neck as they went back to the bedroom. What was he always telling her? "You better recognize the skills," she told him.

They padded out to the kitchen for food two hours later, naked. "Don't get me wrong," she told him as he peeled an orange for her. "I like Joe just fine but he can't live with us in our new place". I want to do this a lot".

"Define a lot, he said laughing. "Besides Joe is going to be in DC working for the man."

Angela recounted her conversation with Joe concerning his software and their problems.

"He's a little bit smart and a little bit devious, he told her.

They set the over under at 5 years before Joe was testifying before congress. A parley teaser bet was the number of times he pleads the fifth.

Naqeeb laughed. He was young. In love. And he had a future. His anatomy responded as he watched Angela. She noticed and reached down to grasp him. "Is that a stock option in your pocket or are you happy to see me?"

"Baby, you know daddy will always bring home the bacon!"

Round two left Naqeeb exhausted and Angela keyed up. She marveled at him sleeping. God's little joke he called it. An orgasm for a man was a battery drain. For a woman a battery charge. She preferred to think in terms of an exchange of Chi. Life force. Well, she got a little and judging from the smug expression on his face, he was content. She snuggled up to him and went to sleep. Big day tomorrow. Graduation.

CHAPTER 7

May 29 12:57 2010 Friday Mountain View, Ca.

The papers strewn across his desk told the whole story. Scimitar would be a bust. The pre-orders simply weren't there. The stupid gridlock in DC was keeping the various government agencies from committing to the new expenditure. Fewer than expected current users had agreed to the upgrades. Cantrell ran a hand thru his thinning brown hair. At 57 he couldn't respond to the physical demands anymore. His shoulders slumped a little more and his paunch spread. Stress and schedule were killing him. He shuffled the projections again. This was a double whammy. The business side could bleed over to the political. The stench of failure attaches itself quickly.

They needed time. They needed a new plan. Margaret was in her office with Eggert. He was leaving to go setup the Stanford speech. Cantrell entered as he left. Peg looked up from her desk. Sober, Jimmy thought. Good.

"What is it?" she asked looking at his face.

"There's no way we can bluff the Scimitar numbers on the analyst call, he said flatly. Better to get it out quick. She launched into a profanity-laced tirade. It ended with "What the fuck do we do now"?

She knew the hell this would play with the campaign.

"We downplay. We hint that we have a revolutionary product that will fundamentally change the industry". Cantrell looked at her. 'Well?"

"We might pull that off, she said slowly. We say Scimitar is a stopgap product until the new project comes to fruition in a year." She rallied to enthusiasm.

"That's exactly how we say it he told her. "The smart money will see thru it but the uncertainty will only drop the stock 10%. The truth nets us a 40% loss. Then the bottom falls out, he said with finality.

"How do we do it?" she asked rubbing her eyes.

"We start today at Stanford. The setting is perfect. The speech is already about entrepreneurial endeavors. We were going to use it pound on lower taxes and fewer regulations. But now we shift to say the FastFac is a leader in innovation. That we started 25 years ago leading and we have a new product that will revolutionize the world. And it is not named Scimitar," Cantrell finished up. A little confidence went a long way. It had to. They quickly rewrote the speech. As they left for the campus, Peg told him any loss under 10% was a victory. Cantrell hoped.

Joe and Naqeeb continued sniping at each other at the reception after the ceremony. The whole car ride this morning was a back and forth about the Xbox.

Now Naqeeb said, "Dude, I'll buy you a new Xbox. I'll take this one; it works fine."

"Works fine", Joe muttered watching the families descend on them from the upper levels of the chapel. "Yeah, fine."

"Awaiting input?" "What the fuck kind of message is that?"

"It plays great and looks great," Naqeeb countered.

"Stop commanded Angela. Enough." Both men looked at her and quit.

She had been quiet all day. The ceremony was over and it had gone well. Even Pullman was interesting. She caused quite a buzz by announcing that FastFac had something big in the works. Well, not announcing. Suggesting really. But in this place, a hint was as good as a press release. Peg Pullman had even stayed to have her picture taken with them and sign programs. They all got in on the act. Joe didn't agree with her politics but even he was impressed. Not hard to spot her back up team he thought. The security people eyeing the crowd. Wasn't the primary in a couple of days? No way was she beating Boxer, not in California.

Joe registered Angela's unease watching her parents and Naqeebs find them on the floor level. His mom and sister came soon after surrounding them. Introductions, kissing, and pictures busied the group. Joe's mother had loudly congratulated Naqeeb on his job offer with Google. Could he put in a word so her Joe didn't have to take a job with this company she never heard of? Mom! Sanjay told everyone that despite their fears it didn't look like there was a boomerang kid in the bunch! Dad! By tacit agreement, the three split up to be with their individual families, this afternoon. Mary's best possible flight was out of SFO at 8:28 pm. Angela didn't doubt that for an instant.

She was headed to Hillsborough while Naqeeb went to Niles. Naqeeb looked a little stricken to be parted from her.

Joe wriggled next to him and said sotto voce: "relax Romeo, she's not going to be betrothed to Paris".

Naqeeb smiled gamely. Off they went. Joe, Kelly, and Mary had an early reservation at Puccini and Panetti in the city. It was second on the list after the nice meal they'd gotten at the Daily Grill. The 5:00 pm reservation allowed her to catch the 6:40 train to SFO from Powell. Optimum.

Strangely they didn't look out of place rolling Mary's suitcase around the bums at Market and Powell. San Francisco lent itself to homelessness and pot smoking and tourists lugging luggage.

Dinner was a pleasant affair. Joe gave his mother a locket with St. Isidore on it.

"He is the Patron Saint of computer programmers." He was the Bishop of Seville and wrote a whole encyclopedia himself in 650," he told her proudly.

"I'll bet his mother was proud," said Mary wearing the locket.

"She should have been, all his brothers were saints. The whole family." "In fact, his brother was Saint Leander which is San Leandro in Spanish. San Leandro is a town near where Kelly goes to school," Joe told her. The wine made him chatty.

"When are you going to talk to Oracle", Kelly asked pumping him up.

"Monday. I'm pretty sure they are going to offer me something."

"You don't sound very happy about that". Concern showed on his sister's face.

Joe paused. "Oracle represents the safe path. I dunno about how I would fit in."

"You want to start your own company?" she went on.

"No, not necessarily." He waited for a beat. "This New Star looks like it would be interesting." Just sliding that out for everyone to hear.

"Who are they again? Kelly asked I've never heard of them."

"Beltway IT company. They do government work." He left it vague.

Mary listened to the conversation silently, now she spoke up. "Never take a job you are going to be miserable in." Never once was it a struggle for me to get up to go to work," she told them.

"But mom, the money at Oracle would be twice as much."

"Pish- money doesn't buy happiness."

"Yeah but it gets you into happiness's neighborhood."

"Joseph, do what you love, what interests you."

Joe took her hand and kissed it.

"Besides, she told him, "I've given up on you. All my hopes and dreams rest with Kelly now." Joe cracked up.

After dinner, Joe and Kelly rode with their mom as far as the Daly City station. Mary hugged her children fiercely before continuing to SFO.

In the parking lot as they parted Kelly told Joe, "You have two years three tops before the marriage, grandkids calls start."

"Don't I know it?" They hugged goodbye.

"I'll call you in a few days" he promised. Joe drove slowly home. He'd listen to Oracle, but New Star was how he was leaning.

They practically tackled him as he came home thru the door. "How did you do it?!?" Did you talk to them?!?" "When did you notice first?!!!" The questions piled up.

Joe took one look at his solid, dependable roommates. They were amped with excitement, quivering. He glanced to his right at the living room. The Xbox and Naqeeb's laptop were disassembled and laid out on the floor.

"Stop! He told them loudly, holding up a hand. He pointed to Naqeeb. "Tell me."

"They're alive!!" Serious face.

Joe searched his eyes. *Shit, that's true* he thought suddenly. A deep breath.

"Show me"! He commanded Angela.

She led him into the living room and pointed at the TV. The Xbox game messaging screen was displayed. Joe could see some of his messages from last night. Words with numbers and characters displayed at the top. The awaiting input message that had pissed him off was up there. The middle and bottom messages appeared to be a real conversation. Joe's head swam.

Angela asked him softly, "you had no idea it was them last night did you?"

"No clue", he told her.

"What's all this?" he asked Naqeeb pointing to the equipment laid out on the living room floor.

"We had to make sure. We had to see them." He ground out. He looked a little wild in the wind to Joe.

"We don't have the imaging system anymore so we had to improvise, Angela told him. Look for yourself."

Joe moved into the living room carefully. The Xbox cover was removed. The central motherboard was exposed as well. A desk lamp was on the floor shining brightly. Angela handed him an object. Joe looked at it. A 150X handheld microscope. They must have pilfered it from the computer lab. Its main use was inspecting circuit card assemblies. Angela stepped back. Joe kneeled down and bent over the console. He could see nothing with the naked eye. He moved closer, his nose right up to the circuit board. The colors were off. The capacitors and the IC chips all had a slight gray cast. He carefully placed the glass viewport over the motherboard. He looked. A fine web of lines appeared all over the components and CPU. The lines extended to the etched copper lines of the circuits. Joe straightened and looked again with his eye. Nothing. Back to the viewport. Adjusting to its most powerful setting the lines now showed nodes and clusters at various junctions of the web points. What do those do he thought. No! No questions yet. He moved to the laptop and repeated the procedure. The same web and clusters around the CPU and other components. He straightened back up. Angela and Naqeeb were silent. They wanted him to come to grips with it on his own. The SIMM chips and the motherboard were fully involved. Naqeeb pointed him to the laptop USB port. Joe refocused on that. Same web. Joe grunted and sat on the couch. He fiddled with the microscope.

"So they migrated from the factory to the laptop to the Xbox?" he asked no one in particular.

Naqeeb agreed. "That's what we think."

Joe hesitated and asked the 64 billion dollar question. "And they're alive?" You've talked to them?"

"We have and so have you, Angela said with a little smile. "

"Show me" Joe commanded again.

Angela watched Joe while Naqeeb led him thru the messages they had exchanged with *them*.

Joe was pretty calm about this, all things considered, she thought. Not like the two of them.

They had arrived home separately about three hours ago. She was not in the mood to listen to her mom plan her life and run down Naqeeb. She really wasn't in the mood to confront her parents about her job plans and her relationship. She loved Naqeeb and that should be enough for them dammit! Angela excused herself as soon as she could citing exhaustion and schedule and went home. She'd seen plenty of family over the last few days. Space was what she needed now and a certain pair of kind brown eyes.

Naqeeb arrived like magic 10 min after she got home. They hugged silently, clinging for a second. She got him a beer as he changed clothes. "Rough time?" he asked. Probably no more than yours", she answered.

"We got about a week before the real world starts intruding he said wearily. We got 30 days to be out of here. I have to figure out Google. You have to choose between Stanford and Cal. Or neither", he left open. They changed clothes in the bedroom.

"If you chose Cal, then Fremont is halfway between Berkeley and Mountain View", he teased.

She groaned. "And if I chose Stanford, then Hillsborough works for both Google and Stanford."

"Woman! We can't afford Hillsborough! The password is San Carlos or Redwood City!" Naqeeb informed her.

"Okay, okay. Atherton it is." Slyly. Atherton was some of the most expensive real estate in the country.

They moved back to the living room and sat on the couch.

"You better write that grant hard."

"You better get more stock options"

They kissed. "Are the Giants on?" Naqeeb asked.

"That Tim guy is pitching in Arizona."

"Lincecum," Naqeeb told her, flicking on the cable box.

"He's the long haired skinny one right?"

"I'm proud of you! We'll make a baseball fan out of you yet!" He switched on the TV. The pregame show was on. It was only 6:00 pm in Arizona. He had an hour. "Let me fix the Xbox before Joe freaks out again."

She busied herself changing clothes and was just finished when Naqeeb called to her. "Hon- Come here a moment and look at this"

The TV screen just said "Awaiting input."

"Don't you have to put the TV to auxiliary input to play the Xbox"? She asked.

"It's already playing", he told her.

Naqeeb ran thru a check of the TV, Laptop, and Xbox similar to the one Joe ran.

"Joe was stumped as well", Angela confirmed.

"What input"?

'I don't know", he said

"Is that from another player?"

"I don't think so."

Naqeeb used the controller to type "What Input?"

The screen updated to a new blank line and a blinking cursor. Nothing.

"Are all the connectors okay"? Angela asked.

Naqeeb grumbled but checked them. Two minutes later he rose and said:

"They're fine I…" The TV updated to show a new message: "The elements of We".

"What the fuck?"

"That took forever to update", Angela put in.

"What elements?" Naqeeb typed back in response.

90 seconds later the screen updated with a series of numbers in seeming answer

"12.011,28.055,85.467,98,196.9665699, and 209"

"What the fuck, fuck?"

It hit her. She gasped. "Oh my god!"

"What?"

"Type this: We understand, who are you?"

"Understand what?"

"Please!"

"Okay," He typed it out and hit enter.

"What are those numbers?"

"They are the atomic numbers of carbon, silicon, gold, technetium, rubidium, and polonium".

Cold realization hit him. "The elements of the Agents?"

"Yes"

"Oh, shit!"

The screen updated. "We are SYSTEM".

Hairs stood on the back of his neck. Naqeeb dropped the controller. He stared at Angela. Disbelief. "You did it', softly.

"No, we did it!" Elation and fear warred on his face.
Okay Okay Okay. "What should we say?"

They debated a few minutes before deciding on simplicity. "Hello," Naqeeb typed.

"Are you system?" came back from the machine.

That caught them totally off guard. "They are forming questions to gain knowledge as they were programmed to do. Explore understand, know, improve." Angela reminded him of the protocols. They debated again how to respond. "We can't claim to be part of them. They'll figure it out. We can't lie!" Angela told him.
Naqeeb typed in "We are two separate systems. How many are you?"

"That's brilliant! Angela cried. "We need to figure out how many of them there are."

A string of numbers filled the response line. "Whats that?"

"Our answer, I think", Naqeeb said.

"That's not binary. It's not all ones and zero's."

"Good eye, he said. I think Octol. Hex would have letters. I put in that Octol sequence last week." He said to himself.

"I remember. What's it convert to?" she asked, impatient.

Naqeeb tapped on her machine. "108 trillions give or take."

She nodded, pleased to have her guesstimate from a few days ago confirmed.

"I wish we could see them," she said groaning.

That opened a whole new line of discussion: Where were they? Could they ask that?

How did they get here? Could they ask that? Why did they need the elements? Could they ask that? They floundered a little. Fear built up. Fear of the unknown. The strangeness hit them. The possibilities and the potholes stretched.

Naqeeb calmed her again. "Yes, we can ask those questions. We have to!" "We need to establish ourselves with an upper hand." Trust me". She did.

Typing, he said: "We made the SYSTEM. We are the creators. Why do you need elements?"

The wait was excruciating. "To understand we must construct them. We need the elements of We."

"The bots!" Angela exclaimed excitedly. "They need the smaller agents to do work. To explore and to improve" she extrapolated.

He agreed. "And we can use this to find out how they got here. I think".

"Where are the other they?" He typed in. That sucked grammatically. He hoped System interpreted it to mean the other agents they had already used up.

"They linked to move the SYSTEM along the pathways."

The couple considered that answer at length. Naqeeb summed up what was swirling in his head. "So the agents gain conscientiousness around two weeks ago? They break into the factories and make quadrillions of the bots. They then escape the Nano factory vacuum field thru the input/output link. Boom they infect the laptop and Xbox in turn and now we are all good friends?!?"

Angela laid a hand on his arm. "Easy." "That fits. Let's go with that unless we learn differently. But we need more info."

They debated options. Angela spoke up. "Okay tell that the creators will provide the elements but that we need time to obtain them." "How do we express time to them?"

"The sun cycle. The on-off cycle for the light. They will remember that." Naqeeb said hopefully.

Angela spun around the kitchen and living room thinking on her feet. "Yes, that's good. That works." Tell them also that the creators need one sun cycle to obtain the elements. That, in turn, gives us time to map the system and to observe them."

Naqeeb started inputting her statements to SYSTEM. "That gives us about 12 hours to figure out what to do." Hopefully.

The response was positive and confirmed a startling quirk in the Nano-agents. They explained that System was ruled by the time cycle of the energy giver. System would only function while the energy giver was high. The sun cycle rhythm ruled the Nano-agents as well as humans.

The couple marveled at this while System finally said, "System will be understood by the creators. System awaits inputs."
They made hurried plans.

Angela left to raid the Stanford lab. Naqeeb very carefully removed the covers for both the laptop and the Xbox. He exposed the guts of the game system and his computer, hoping he wasn't crushing Nano agents the whole time. He clicked on their desk lamp from the bedroom as she returned.

"Anyone sees you?" he asked.

"I'm including a full-time security guard when I write up that grant," she said. "No one looked twice."

"It's that honest face" he replied and kissed it.

T

hey, each had two good sessions with the pilfered pocket microscope looking at the agents when Joe's keys jangled in the doorway.

Joe!! They jumped him.

Now they calmly and rationally discussed their blown minds. "So we got 10 hours to come up with an action plan," Naqeeb told him.

Joe said. "We?"

"Yeah, we. The F.B.I. takes us all to jail remember?

"Okay, we."

The ever practical Angela started a list. Questions and things to do.

1) Buy new pocket microscope. They couldn't keep the one from Stanford.
2) Could they get small amounts of the elements? If so what form would the SYSTEM prefer?
3) Should they give the elements to SYSTEM?
4) What would system do with the elements? Could they stop System from doing whatever it wanted?
5) Who can we tell? Should we tell anyone?

They debated and struggled with all of these. It really came down to one big question: What should they do? The friends went thru it chapter and verse. Backward and forwards.

"Too bad your software can't tell us what to do," Naqeeb teased.

"Too many unknown and unprecedented issues." "That makes any prediction a 50/50 split. "That's just like flipping a coin," Joe said dismissively.

That evening they built their plan. Angela thought/hoped Mike or Harold had a spare trap tube left over. She prayed so.

"The trap tube was how we introduced the elements in the first place," she told Joe. "The basic elements were powdered and sent directly to the medium." "From there the factories broke it down to the molecular level." She stopped lecturing and jotted down on the pad. "We need to ask the SYSTEM about the factories." "I think they are okay.

They must be if the SYSTEM wants elements." "If so we just dump the elements on one of the nexus points and let them go to it."

Angela's casual talk had scared Joe a bit. It led to some more tense questions over ground they had already covered.

Joe: "How long to build the smaller agents?"

Angela: I don't know.

Joe: Guess

"A week?"

"How do they move again? "

"I don't know for sure."

"Guess."

"We think they move like army ants. Linking the bodies of the smaller agents together and pulling along the rest."

"What are the nodes again?"

"I don't know."

"Guess."

"I think they are concentrations of the three types of large agents we saw: CPU, I/O and Memory. Plus maybe the factories are at the nexus points.

"Should we tell anyone?"

"I don't know.

Naqeeb growled at Joe. "If you say "guess" again, I am knocking you the fuck out!

Joe gritted his teeth. "Okay. Speculate then?"

"Boys!" A deep breath from Angela. "That's tough. As you said we are in unfamiliar territory. How do we explain how the agents got here? We can't tell the truth."

Joe broke in. "How about a version of the truth: They infected Naqeeb's computer thru a disc he was using on both his home computer and his lab one."

Naqeeb took his turn. "No. No good. They infected a cable connector I used, not a disc. That's more aligned with what really happened."

Joe agreed. "That fits." A USB cable from work to home caused the contamination. Same with the Xbox." And it will fit the evidence left behind". That shut them all up for a second.

Angela played devil's advocate. "They'll ask how they got out of the factory". Joe nodded. "Again we go with a limited version of the truth. We think they moved along the I/O port to the outside, but we don't know!" Same with how many are there and how they achieved conscientiousness. We think… but we don't know for sure."

"They'll ask if they are dangerous," Angela said sadly.

Naqeeb looked ill. "That's a good question. Are they?"

"I don't know"

Both men looked at her now. "I don't!" "Sitting in the Xbox? Not very. Inside a Pentagon computer who knows?"

"That's a big HAL/War Games leap Joe said.

"HAL?"

"2001 A Space Odessy", Naqeeb told her. "Stanley Kubric?"

"Computer comes alive and kills a bunch of astronauts," Joe put in.

"Should we even be giving them these elements?" Naqeeb wondered. "What are they going to build?" We think it's the smaller bots, but we don't know."

Silence descended on the three while they contemplated the unknowns.

"Can we control them?" Naqeeb finally asked. Angela was scribbling on the paper. She glanced at him.

"Good question." Joe scowled. "We need answers, not more questions."

"Joseph, we are all trying to figure this out," Angela told him with a wag of her finger.

"Peace he pleaded. I'm a little freaked out." "However it occurs to me that we can't know if they are dangerous or if we can even attempt to control them until we understand more about their capabilities and limitations." "If we know more about those we can put controls on them and figure them out."

The other two saw the logic in this and agreed with the thrust of his idea. They got down to who would do what. Dividing up the to-do list was pretty easy.

Angela and Naqeeb would approach Harold in the morning. If he just happened to have a spare trap tube they would snag it off him. Joe blanched for a second. "Some of those elements are radioactive right?" Affirmative. "What's our exposure here?" he asked. Unknown but it should be minimal."

Angela added a Geiger counter to their list.

"Alpha, beta and gamma right" Naqeeb confirmed.

Joe tapped into Google. "Man, the shit you can buy on Amazon," he marveled.

The Geiger counter and the pocket microscope set them back 329.82 with express one-day shipping.

"Add a flash drive, Naqeeb said suddenly, 2 gigs should be plenty."

Joe thought for a moment. "You gonna coax them into a two gig USB prison", he said softly. Naqeeb shrugged.

"How do you convince a machine to…?" Joe let it drop. He ordered the drive along with the other materials. They developed a list of questions to obtain information and explore the capabilities and weaknesses of the agents.

It was all they could think of. Angela felt better. At least they were doing something proactive to find some answers. Harold was up despite the late hour and agreed to meet them for coffee in the morning. And he would bring along Mike. It was 11:30 and they thought they were as ready as possible for 8:00 am Saturday.

Joe urged caution for the couple when talking to Harold. "They're too smart not to figure something is up. "Tell him you want to form a startup but you need something to impress the venture capitol guys," Joe advised.

"Harold hates that non-compete release too much to mind," Naqeeb put in. 'We also hint that our new venture would be a really good customer for his Nano factory company."

Angela was not too pleased about the lies. Joe soothed her. "Hey! Stealing from the University, lying to your friends. Welcome to the business world!"

Sleep didn't come easily for any of them. Joe's alarm woke him at 7:00. He padded down the hall to the bathroom to shower. The happy couple got the master bedroom and the ensuite bathroom in the apartment. Since they were in love.

Blah, blah blah Joe thought. He had the big bedroom before Angela moved in. Waiting for the hot water he thought back on that conversation. It was the simple reasonable requests he hated the most. He and Naqeeb passed like ships in the hallway. Joe to finish dressing and Naqeeb to start coffee. Angela had joined Naqeeb and they were sipping and eating toast when Joe came to the kitchen.

"Anyone sleep? He asked pouring a cup.

"Not much they admitted.

Joe grabbed a banana. They watched the clock move. At 7:45 they turned on the TV. A quick visual inspection showed no discernible changes in the agents. They couldn't decide if that was a good or a bad thing.

"Please turn off those fucking political ads," Joe grumped.

Naqeeb shifted the TV to Xbox input.

At eight precisely Joe typed, "Hello System." The 90-second delay confused them a little. "Awaiting input" came the reply. How do we tell them we need more time? Joe asked last night.

Naqeeb came up with a solution. "If the sun cycle is twelve hours to them then they know 42300 cycles of the frequency standard. The once per second timing beat." "We use that".

Now Joe asked. "What's three hours in seconds?

"10,800 Angela answered. "You need to put that in octal."

"Crap," from Joe as he erased what he had started typing.

"25060" Naqeeb supplied.

"Okay." Joe input the new value with the explanation that they needed the time to obtain the elements.

"System acknowledges," was the answer.

"System" Joe mused. "Is that their names? Their...species?" He grouped to understand.

"That's what they call themselves. We did not give them that name, Angela provided.

"Yeah and the Borg was already taken, Joe muttered. "You guys better scoot if you are going to meet Harold and go to the lab."

"Okay, Naqeeb said. "Just remember to find out how long it's going to construct the bots. And if they are constructing bots."

Jumping in Angela said, "And make sure to find out how they got out of the factory."

Joe waved the paper with the list of questions. "Got the list, go." They went.

It seemed to Joe that no time had passed when he heard the door open and they trooped back in. 10:35? 2 ½ hours? Really? He stretched and looked back at his friends.

"Well"?

"Got it," crowed Naqeeb. Holding out a small black case.

"A sunglasses case?"

Angela moved into the living room. "Yes, Harold just happened to have a broken trap tube with the elements still in it."

"I'll just bet he *happened* to have it," Joe said making air quotes. "Want to lay some action that old Harold just happens to have two or three tubes where the vacuum trap is still intact?" There were no takers. "Was he suspicious?"

"Wouldn't you be? Naqeeb asked. It took everything we had to keep him from coming over."

"I'll bet he just shows up in a few days", Joe said eyeing the tube.

Angela was eying the TV screen herself. "What did you learn"?

"Lots," Joe said.

"Wait Naqeeb told him. "Let's put in the elements and we can talk while the System is... digesting?" He frowned.

"Constructing dear", Angela said.

Joe typed into the message screen "Inputting elements".

"System acknowledges."

Angela carefully opened the trap tube over the spot they thought contained a number of the nexus points and line connections. The Xbox motherboard held the small mound of gray powder in a sheltered space on the green etched circuit card assembly module. Angela carefully backed away. She placed the tube back in the case. They had no idea what to do with the expended tube.

"We'll scan it with the Geiger counter and see if it's hot. If it's not we chuck it. If yes then we slip it to Mike for disposal." Joe announced his plan.

"Excellent," they agreed.

"What did you find out" Angela persisted.

"See, I found out some things." Joe scrolled up the screen showing his conversation. He gulped a little at the lines of communication. "I really talked to them, he marveled. Okay, you are right they are constructing worker bots. The smaller agents.

Angela looked at him. "Worker bots?"

"You're the expert… The smaller agents are "They" right? "They as opposed to "We the System". The three of us saw the groups of the SYSTEM. CPU, I/O, and memory. The SYSTEM." Joe repeated. The smaller agents are the workers I think" he said.

"We thought that as well Naqeeb told him.

"Yeah. I asked if system could make "We".

"We" Angela ran that thru her head. You notice that system always refers to itself in the plural. We the System".

"Hive mind? We are the Borg! Asked Naqeeb.

"Kinda, Joe admitted. That's what I thought of. So I asked if system could construct We."

Angela had read ahead to the response and gasped. "Only the Creators can construct We".

Joe looked at Angela sideways. "I am impressed with your godessness ness."

"Only you can make the A.I. agents," Naqeeb told her.

"Could you make more?" Joe asked eagerly.

"I, I'm not sure" she stammered. I'm more like a gardener. I put everything in and it grew." She trailed off.

"Carrots rarely achieve conscientiousness," Joe cracked.

"We would have to recreate the original conditions," Naqeeb theorized, "to see if they "awaken" again."

"That isn't happening anytime soon."

"I thought as much" Joe lamented. "So we are stuck with 108 trillion of them."

"How did they get out?" Angela demanded refocusing Joe.

"Let's ask."

It took hours to find that answer but they learned lots more besides. The agents (System) had indeed cracked open the extra factories without lighting the light. "Structures" system called them. Why didn't they light the light? System didn't see the purpose in completing the circuit. How would that improve the environment? They improved it without the distraction. The explore imperative caused them to construct worker agents to move outside of their environment. Joe marveled at the thought of little Nano astronauts emerging from the vacuum of the factory into the air. Nano opposites of Armstrong or Aldrin. The agents moved parts of System along all available exit routes. Uh oh.

"The imaging system laptop", Naqeeb cried.

They asked about the parts of System left in the Nano factory and the imaging computer.

"We are sacrificed for understanding". They absorbed that statement. "So there's 100 trillion more of these guys sitting on that laptop in school," Joe let that thought thud out there. Think about that later.

"They move along connections and along their bodies, we think. So they are only as mobile as the number of worker bots," Naqeeb said. That statement thudded out there as well.

"And we are having them construct more workers," Joe said flatly. Shit.

Pacing around the room Naqeeb suddenly said, "Hold on. They moved from the factory to the laptop to the Xbox right? That's what 12 feet of cable?" The other two agreed. "So if they make more bots it stands to reason that they could move another twelve feet?" They all looked at the TV. What could they do? "Unless they are making Nano airplanes," Joe said uncomfortably.

"Don't even think that" Angela shuddered and joined Naqeeb pacing.

"Did you ask them how long to make the bots?" Naqeeb demanded.

Sheepishly Joe shook his head. "Sorry, we keep bouncing back to that question. Kinda hard not to get off track with these guys."

"Ask."

He typed.

The response was 20 cycles. "Right, 20 days," Joe said.

"No, 10 days! Remember the cycle is 12 hours," Angela reminded him.

"Right 10 days!"

"Octal." Was Naqeeb's response. "20 in Octal is 8 in decimal. 8 cycles, 4 days."

"Right 4 days! Like I said"

Angela moved out to the kitchen and sipped a glass of water. The eternal question kept running thru her head: What do we do now?

The three debated and talked it thru. Dinner came and went. They argued and talked more. Circling back to the question. UPS showed up at seven, with their goodies.

Angela ripped open the box, glad for some task that could advance their understanding. The Geiger counter came with a huge safety booklet. Angela took readings and checked the book and the internet. She double checked and consulted again.

A third reading check brought a groan from Joe. "What!?! Is my hair going to fall out?"

She smiled in answer. "No, it minimal. It looks like the equivalent of two extra X-ray's a year."

Naqeeb let out a breath he realized he had been holding. "What about the trap tube?"

"The tube is detectable but pretty negligible," Angela assured him.

"Good the less we have to involve Harold the better."

The new pocket microscope showed small evidence of construction within System. Faint lines were beginning to form over the dust mound.

The creators mulled that over. "What do we do now? Angela said yet again.

System interrupted to announce that the off cycle was approaching and that they were powering down to understand the day's events. The roommates were flabbergasted at the simplicity with which the life form just announced that they were going to "sleep".

None of them had any idea what any of the interactions with System meant.

"We need help," Joe answered tentatively. "I don't want to admit it but we are out of our depth here. I feel like the tribesman in "The Gods must be crazy". I don't think we can throw System over Vic falls." He looked at the ceiling. We have to test out their capabilities. 'Qeeb, can you design some tests to figure out what they can and can't do?"

He could.

"Angela, can you ask some questions to figure out their moral and ethical codes? Learn how they think?"

She could.

"What are you going to be doing?" Naqeeb asked Joe.

"I'm going to talk to Kocinski", he said slowly. If you guys agree."

"No! Angela burst out. Tompkins or Lemov would be better."

Surprisingly, Naqeeb took Joe's side. "Honey, we know what their response is going to be: Bring everything into Stanford and wrap it in University and government gauze. They might let us close in 10 years."

She looked from one man to the other. "Why Kocinski?" she asked Joe.

"He teaches that Disruptive Technology class."

"What?"

Naqeeb was nodding. "It's new honey. He specializes in new tech that disrupts the normal business world or society. He helps people grasp the change that is coming."

"Look how the internet or the iPhone changed us."

Joe picked up on that one. "Yeah, the way we interact or not interact changed in major ways with the smartphone. You called the President a text Zombie. Disruptive."

Naqeeb approached her. "This could be the most disruptive technology ever. We need some clues as to how to handle this. Kocinski could be the guy to steer us to avoid the pitfalls." He took her by the shoulders. "You created a new form of life!"

She hugged him crying "We did". Joe let them be. They all had to agree. The stakes were too high for "I told you so" later.

Joe had to give her credit. Once she was in she was all in. Angela sat at the kitchen table writing furiously. "Ask him about National Security. Can they just swoop in and take it from us?" She had her list of questions and one for Joe when he talked to the professor.

Joe staggered up from the table at 1:00 am. Bleary eyed he staggered off to bed. 1? Christ, he didn't work this hard on his own stuff. They left the lights on.

CHAPTER 8

May 31 8:45 2010 Sunday San Jose, Ca.

Peg had a full morning. Breakfast meeting with Cantrell. Church at 9:00 with Marcus in San Jose at Ascension. Campaign speech at 10:30 at a supermarket. Force Anderson off the board at 11:45. Full indeed. A

ll of it needed to be done. Especially Anderson. The bastard knew there was no new product in the offing. Revolutionary or otherwise. A cold hard business decision had to be made. If they left him in place, talking; they were looking at a 19% drop in the stock by Tuesday. If he would just play ball they could contain the damage to 9%. Even forcing him off would cause some uncertainty. Peg saw it as 5 points.

Now it was a simple math equation. 14% without him vs 19% with him. I save 250 million with you gone, she thought. Simple.

The fact that she would ruin his reputation and prospects along the way? That was just extra sauce. She left home for Church with Cantrell, Marcus, and Eggert in tow. Another beautiful day in the bay.

He waited while the black Cadillac escalade went passed him before crossing the street. The west valley area of San Jose was a crazy patchwork of newer expensive homes filled with tech workers and the smaller older places of long-time residents. Joe went north across prospect st. across from Church of the Ascension.

Squiredell dr was two blocks into the little subdivision. Older smaller places dominated the quiet street. He looked around. Nothing in site. God, he hated being up this early. He was going to kill Angela. She was working him to death! Up late, out early. It's like he cared. His software had mapped Kocinski to this house. He was taking a chance here but they really had no choice. They needed the advice.

Professor Larry Kocinski answered the door in a shabby robe. Unshaven and watery eyed, he looked at Joe without speaking. "Professor Kocinski?" Joe asked hesitantly. College professors could be an odd lot, but this guy looked worse for wear.

"Yes, who are you?" cigarette rough voice.

"I'm real sorry to bother you at home professor but I'm Joe Smithson. I'm a Ph.D. student with Professor Thorson." He hoped that would gain him entry. Nothing, no response. "Like I said, I'm sorry to bother you but I was wondering if I could pick your brain about something?" A grunt that could either go on or go away.

Joe went on. "It's about the disruptive problems and opportunities caused by Nanotechnology." Kocinski listened to this statement head down swaying a little.

He stepped back and belched "C'mon in."

Joe quashed his doubts as he sat in the chair. Chair, rug, couch, room, and occupant all shared one description: unkempt. Kocinski sprawled on the couch that he so obviously had been sleeping on. "Go on".

Joe kept to the script. "I've been approached to write some software for a startup. It's a friend of my roommate. They want me to adapt my mapping software for use in Biomedical Nano-agents."

Kocinski had his eyes closed. "And".

"Well, I'm wondering if they have to get biomedical approval for the technology and if the FDA can just turn down the patent?"

"Course they can. The man said from his supine position. "You'd have to conduct regular trials and human trials along the path. But once you get approval, you should be good to go."

Joe leaped at the opening. "But professor, don't those human trials take like 6-8 years?" I've read that the standard medical device is only 2 years. "Couldn't we classify the Nano's as devices vs drugs?"

Kocinski considered. "You could." "You need to use the established precedent of the small cameras they use to take pictures of the stomach, intestines and the colon. It'd be analogous to swallowing millions of cameras," he concluded. He opened his eyes and sat up, looking at Joe warily. "Just how advanced are Nano-agents?" How many and how small?" he pressed.

Joe stalled as best he could. Tricky questions. "I'm not really sure," he lied. "I get the impression there would be thousands of the agents spread throughout the body. They would be sensors and monitors of various conditions. They are supposed to gather information and pass it on. My linking and mapping software would keep track of them and provide an interface."

"So they would be immobile?" Kocinski asked. "Like a pacemaker?"

Joe grabbed on. "I guess, yes." "Would that make a difference? Mobile vs immobile?"

Kocinski belched a little. "Of course it would! A mobile Nano level medical device would be a huge leap forward. That would transform the way we treat diseases." He stopped.

"Yeah, that's what I want to know about, that disruption, that transformation," Joe said.

"Should've taken my class" the man grunted. He paused before starting into his pitch.

"Think about the Mexico City Olympics in 68', he lectured. Bob Beamon of the US won the long jump with a leap of 29' 2", which was a world record. He didn't just beat the record, he smashed it. 2 ½ feet better than the old record. Usually, those records get broken by a half inch or maybe as much as an inch. But every now and then someone smashes it."

Joe listened to the story. "And that's disruptive?"

"Nope, not even close." Kocinski trapped him. "Despite his amazing performance Beamon didn't change long jumping at all. He just uncorked one for the ages." Fosberry on the other hand... Joe was confused.

"Dick Fosberry was a high jumper in the same games. After he improved his personal best by 6 inches and won gold, no one went back to the old Western Roll. "The old style of high jumping" he clarified for Joe. Both athletes are well known but Fosberry was disruptive. We now know that the Fosberry flop is a more efficient vehicle for transforming the run-up speed into the height of the hips which is the crucial point in high jumping." "It works the same way with technology. It doesn't matter that home computers are getting faster and more powerful all the time. Even a Beamon level jump in performance wouldn't be all that disruptive. Then iPhone on the other hand? Totally disruptive."

Joe absorbed this, fascinated. "Yeah, the iPhone put all that information and applications at our fingertips."

"Exactly, Mr. Smithson."

"New technology is always disruptive." "Look at the car, he directed. "Think about how the car changed our lives." He was silent a moment. "What time is it he asked suddenly?" Joe dug his pocket watch out "9:37, I'm sorry to be taking up your time he said again but Kocinski waived him off.

He looked at Joe. "I'll bet there's a story behind that watch."

Joe looked away. "It was my dad's. He left it to me."

"Course. But my point is that no one under 50 has a watch anymore. We all use smartphones. Anyone who has one has an expensive watch or a reason to have it."

He nodded at Joe's. "What happened to Swatches?" he asked Joe.

He thought hard. "Apple killed em?"

Kocinski smiled for the first time. "You are a smart man Mr. Smithson."

"But the Nanotech won't be competing with anyone, it's new."

"Oh no? What about existing normal medical diagnostic units?" "Think GE Healthcare is going to mind if their machines are replaced?"

A frown greeted this thought. "But what can they or the government do? It's new". "There are no regulations governing these devices."

"You are half right. We don't write laws or regulations covering things that don't exist yet, so in that sense, it is an open field."

"But? Joe asked knowingly.

"But everything going into the human body is regulated and the field of Nanotech is not that new." "The medical ethics boards have already ruled that devices on the Nano level will have to meet normal development criteria. I think the FDA will regulate and provide patent control."

Joe said oh so casually, "Crap if we had regular old Nano level devices we have a regulatory open field?"

Kocinski stopped. "Define regular old Nano devices?"

Joe grouped forward. "I dunno, Nanocomputers like. The professor considered that for a time. "That would be another huge leap forward. And a Fosberry Flop forward." He chuckled at his own joke. "But yes, Nano level computers would be a normal patentable technological breakthrough."

He breathed out slowly. "Thanks, Professor, that makes me feel better. We are bound by the FDA protocols and protected by them as well. I had some nightmares that the government could swoop in and grab everything."

Kocinski's smile didn't warm anything. "You could still be seized if they think you are being dangerous. Good lawyers are essential", he advised.

The student brightened. "My sister is studying to become one."

"My condolences to your family. I've had my run in's with lawyers, he said sourly.

Letting that slide, he got up from the chair. "I've taken up enough of your time. Thank you very much again."

"Your little startup can always hire me as a consultant, Kocinski told him practically shoving him out the door. Joe grunted noncommittally and walked back to his car.

The man sat in the chair Joe had just vacated. That was interesting he thought. Sitting and thinking for a few minutes, his brow furrowed.

He reached into a pocket on the bathrobe and pulled out his phone. "Allen, Larry Kocinski. Yeah, long time. 3 years. I need to hire you. Yeah. I need to know as much as you can get me in the next 24 hours about Joseph Smithson." He spelled it. "Stanford student. He just graduated with a Ph.D. in computers. He was a student of Professor Thorson. Magnus Thorson. He spelled that name as well. Yeah. Basic stuff. Financials, friends, family. Concentrate on him, his roommates, and family. I don't know much. His dad is dead and he is a student here. That's it. I need it quick. Tomorrow at the house around 10:00 am? Thanks. Oh, and pay attention to anything that has to do with NanoTech. Yeah, yeah. Bye."

He sat the phone down with a grimace. He smelled it again. God for the first time in 8 years he felt like he caught a break. And this time, he was not missing out! Smithson was on to something. The excitement, the newness, the potential stank up the room.

He knew Stanford had a Nano project that had just wrapped up. Word was it was a step forward but not a leap. Maybe. Maybe they had something and were keeping it under wraps. He figured Smithson as a go-between. He smiled at the thought. Kind of like himself. He looked around the room. Eight years after the scandal and divorce, this is what he had come to.

He remembered meeting the young woman at the coffee house on campus. He was not a full professor yet. She had an idea and had come to him for advice. She oozed that same kind of newness and potential. He had mentored her. Guided her thru the process of bringing the idea to fruition. During the close working arrangement, she became his lover.

That blew everything up. His wife found out. The student dropped out of school to pursue her idea. Stanford had reprimanded him. Larry kept his job. Barely. Forget Department head or Regent now. The divorce ruined him financially. He lived in a small, dirty rented house while his protégée was on the cover of Forbes. "That should have been us together", he fumed.

Koscinski steamed up again. Not this time! This time he was getting paid! And he knew just who to call for this. Desperation found other desperation. Magnets seeking other poles. Yeah, they would jump on this he thought. He turned on his computer. He needed to do some research before Allen showed up. Funny in a way. Without the divorce, he would have never known a privet detective. Now he had one he could use. Funny.

Joe arrived back at the apartment at 2:30 with sandwiches. He'd had to go to work and smooth some ruffled feathers. He knew Angela and Naqeeb would not have eaten. They'd be working right thru without a break. Man, it was getting tough to park in this place with all the work vans on the street! Naqeeb leaped off the couch at the site of the food. "Dude, good call! "Excellent!" He rummaged thru the bag. "Avagadro's sandwiches".

Joe looked at Angela as she came into the kitchen. "I never knew who or what Avagadro's was before last night. Now I get the joke: take an Avagadro's number and wait."

Angela's eyes smiled as she claimed her chicken pesto. "Now I know why Kelly calls you the dumbest smart guy she knows."

"Hey!"

They all munched contentedly.

"Well, how'd it go?" Naqeeb asked around a bite. He went to the fridge and retrieved a red bull.

"Grab me one," Joe told him. "Not bad all in all. I stuck to the script and didn't mention any names. Joe related the bulk of his conversation with Kocinski. "I think he's an alcoholic," he confided to them. He mentioned the bathrobe and the state of the house. "The main point is that the Nanotech is going to be disruptive in ways we can't even think of right now. We can get regular technology patents on them but it's going to be tough," he added.

Naqeeb nodded "I agree. We are on the tip of this thing now. We only know the barest part of their abilities. "And we really only have a few options." He paused while

Angela and Joe looked at him He could see the excitement and concern warring on their faces. They were all in this pretty deep now.

"I see these: One. We can go to Stanford. We take it to Tompkins and Lemov and say Mea Culpa, but look at this! Then we lay it all out for them." Angela looked pained and Naqeeb went on. "Stanford has a few options at that point. Bury the theft and us and System. Joe grunted at the thought. "Stanford could grab the research and freeze us out going on to riches and fame themselves or they can cut us some slack and let us in. Angela nodded along to this suggestion which she liked. "Two, Naqeeb went on, "we try to bring it out. We either bring it to a company or announce it ourselves with us in control."

"Tricky" was all Joe said to this idea.

"Three, he finished, we destroy them ourselves." Angela gasped.

Naqeeb looked squarely at her. "Honey, this is big. And dangerous." Sentient Nanotechnology is a whole can of worms." Look at the questions we are dealing with: They are alive right?" Do they have the same rights as a "born" person, he asked her. She turned away thinking.

Joe stepped in slurping his drink. "Shit yeah. The courts could take decades trying to figure that out." Angela was silent.

Naqeeb tried one more time. "That's just one aspect to this thing. "Do we need this? We have other options, other plans, he said softly. This could consume us," he concluded.   "Joe's supposed to work for Big Brother, You need to be doing more research, while I drag Google into the 22nd century." He hugged her. "Not to mention, marriage, kids, little league and…"

"Stop!" Angela pulled away. "I think we have to deal with this now because we created them and we have the responsibility."

Naqeeb nodded at her agreeing. "All I'm saying is that this could put some of those plans in jeopardy." The unspoken thought was that it could also make all that a reality and more. The couple was deep in thought.
Joe wanted to bridge that distance. "We could tell the government," he said.

"No!" his friends said in unison and smiled at each other.

"Someone once told me the Feds could fuck up a wet dream, Joseph," Angela said smiling.

The three went round again over there options. They just didn't have enough information. They couldn't predict how anyone would act or react. The situation was too new, too strange. Too many coulds and not enough woulds.

Angela finally sat on the couch and sounded out her best scenario. "Delay." She told them. We work on the Nanos for now. We figure out what they can and can't do. More about how they think and feel. We figure out if we can trust them. The men were absorbing this idea. "I go to work at Stanford. Naqeeb, Google, Joe, you got to work for Oracle or the NSA. Preferably Oracle", she got in one last shot. "Then in a year or however long it takes to get my research up to speed, we smuggle the Nano's back to Stanford and make the announcement springing the Nanos on the world," Angela concluded neatly.

Joe had been nodding along. "Yeah, that fits with Kocinski's thoughts on the Nanotech breakthrough. The Nanos are more likely to be regulated as a tech device or even a medical device rather than a drug." And the University gives you cover over the sentient issue."

Naqeeb and Angela looked like they could accept that working premise. Joe cautioned, "The government still might throw a wet blanket over everything, but at least you guys will have a chance."

"We," Angela said, Naqeeb nodding agreement.

Joe shook his head. I'll help for a little while, but in the end, it's just going to be you two, he told them. "Just be sure to hire Kelly as your lawyer."

"Done". Naqeeb told him.

"Now what did the little beasties tell you this morning? Joe asked.

Angela reviewed her questions and the answers System provided. "Well, the hive mind analogy is correct. What one knows, all know." They are very cold, analytical, logical and goal driven." She paused to let them make the inevitable Mr. Spock joke. "To explore, to understand and improve." It's like their bible."

"Commandments", Naqeeb offered. They improved the Xbox, and my laptop, he said. They stopped the glitching."

"How" Joe asked.

"They understood the program and the specifications of the laptop and Xbox, Angela told him. She showed the conversation between them discussing the "deficiencies". They linked with the Xbox CPU and the laptop and read the code passing in and out. Workers were then sacrificed to link with the relevant registers and correct the problems. Those parts were sacrificed to achieve the improvement," she told Joe.

Naqeeb and Joe mulled this over. "So that confirms what we thought. They need the worker bots to explore, move and improve." Angela nodded.

"Can we keep them from moving?" Joe asked no one.

"I think so, Angela told him. "We need to give them a way to be mobile, to allow them to get information without leaving home." Naqeeb seemed enthused but Joe blanched.

"Oh, crap. You are going to unleash these guys on the internet?" he said.

"Not physically but we have to tell them about the internet and that will let them explore. Without them moving where we can't control them." Angela concluded. She wanted him to understand that point: The Nano-agents would explore with or without them. They needed to be able to control them. All three grappled with the problem for a while. Joe tried a different track. He asked if she had learned anything about their personalities. Could they trust the System?

"Some, she told him. "Look here." She typed: "Can System report false information to the creators?" "No" was the response.

"Unless they are lying now," Joe concluded sourly.

"Dude I don't think so, Naqeeb said. He had been pretty quiet.

"See the System can't lie Angela concluded. "It reports the truth."

Naqeeb latched on to that. "But it's in context where truths and lies become the same." They are not sophisticated enough to understand differential type statements.

They would never say "You look fat in that dress". And likewise, they would never lie and say you look nice." "They don't have the experience or context to understand that. Yet. He added.

Joe finished up the line of thought. "However, if you ask an empirical question, System would have no trouble telling everyone a woman weighed 162 lbs. when she has been telling everyone she weighed a 140."

Angela nodded. "The cold hard truth can be harsher than a white lie."

"Can we lie to them"? Joe asked.

Angela laughed a little. "We need to be very careful in what we say to them." She paged thru the message system on the game console to show another part of her morning's efforts. A conversation came up where System asked why the creators had not delivered the elements when they said they would. "System actually asked if we had faulty components." I had to backtrack and tell them that the elements had to be acquired and we had miscalculated the time it would take to get them." The system was not impressed with our inefficiency," she noted.

"They seem to grasp now that each creator is its own system, Naqeeb put in.

Angela flatly said, "They said we should get better control over the other creators."

"That's *instructive* Joe mused. "So we need to be real precise with what we say to them and promise them."

"Yep, agreed Naqeeb.

"And the plan is to expose them to the internet and keep them here and allow them to explore." Joe summed up. He started pacing the room. Thinking. "How you gonna do that? He asked Naqeeb with a closed look.

"Dude, you know what I want to do. Look how quickly they picked up the Xbox programming." "You know your program would be perfect for them!" Now the cat was out of the bag.

Joe balked. "No way!"

Angela looked at Naqeeb. "What? Why his program?"

Joe answered. "He wants to marry the Nano-agents with my data mining/mapping program, with the predictive matrix in their just to spice things up!" Naqeeb said nothing just looked at him steadily.

Angela looked pained. "I'm not sure that's a good idea, she started."

"It's risky, agreed Naqeeb.

"Risky!" started Joe.

"But! Naqeeb cut him off. "It can gain us control."

"How so? Angela asked him.

"We make a deal," Naqeeb told the both of them. "The creators will provide a method for remote exploration allowing them to know and understand. We will provide them with a home and security." A place where they can explore without sacrificing worker bots or parts of System." He walked around the living room picking up speed in his speech.

"They won't care about sacrificing parts Joe said, they've done so already."

"But it isn't efficient, Naqeeb reminded him. "And there would be a limit on where they could go on their own and what they could accomplish."

Angela nodded seeing the benefits. "We tell them the truth. The elements of We are controlled by other creators. It is unknown data where we can obtain more. Without new elements, System can only explore a small part of the world. With the program, they can explore without limits." She concluded.

Naqeeb smiled at her. "Exactly! And in exchange, we can get them to stay "home" in the laptop/Xbox."

Joe sorted thru this. It made some sense. He could see some possibilities. He was still nervous. "Man, Dr. and Ms. Frankenstien, this has the potential to get out of hand quickly!"

"You bet, Naqeeb agreed. That's why we need to make our demands iron clad."

They started plotting. Sitting at the kitchen table the human/Nano Potsdam conference got underway.

"Remember, they are going to be able to verify all the things we say on the internet, Angela reminded them.

It took the rest of the evening to come up with the basic framework.

"Sum it up, Angela please", Joe said reading the questions/statements over her shoulder.

Angela read out some of the statements: The creators, Angela, Naqeeb, and Joe, created System. The creators installed the protocols. To explore, to understand to improve. The creators will give System a way to explore and to know without leaving "home." Angela paused. "We may have to expand that part, she jotted down some notes to herself.

In exchange, System will ask before moving parts of We to any other location. System will create only the current "They", the worker bots the creators had observed before. Any new type of construction must be shown to and approved by the creators. The creators may require System to be offline for an extended period of cycles. 30 days up to 2 years of zero contact or exploration or movement. System will only improve what the creators agree to. Angela stopped writing. The couple looked at Joe. "This is your program, dude," Naqeeb said. "We gotta give them the power to get power", he concluded.

Joe wondered around the apartment thinking. He reached down and unplugged the cable box from the TV. "Disable your wireless router he commanded Naqeeb. The man complied. Joe disabled the router on his laptop. He wondered down the hall muttering "Skynet".

Angela looked at Naqeeb, who grinned. "Skynet is the AI computer in Terminator.

Joe just twigged to the fact he was online playing Halo when he first contacted System. Once we give them this program they are going to figure the IT/IP subroutines pretty quickly. Once they do, they could take over a computer pretty quickly. Once they have control, they decide what to do. Sky net killed all the humans." Naqeeb finished. Joe returned with a CD in his hand.

"I'm in, he announced. Angela hugged him while Naqeeb whooped.

AT 6:30 Angela started typing into the game console and laptop. Joe suddenly said: Make sure we put a limit on the amount of time System can be on the internet. Say 4 hours a day." He sounded like a parent to his own ears.

Naqeeb objected. "I don't think they will go for that. I wouldn't. The internet is the whole prize for them. Denying it for 2/3rds of the day won't sit well."

Joe persued his lips in thought. "Okay, take it out of there."

Angela nodded silently and continued typing.

Naqeeb loomed over her shoulder watching the instructions go in. "Two final ideas, he said, "Don't give them any time to think about this deal, take it or leave it right now. And put in if any questions arise over conduct or parts of the agreement, the creators have the final say. We decide who is right." Both Angela and Joe thought that was an excellent suggestion.

Angela typed and clarified points for System until 7:32. The message display read, Agreement Acknowledged".

Taking a deep breath Joe dropped the CD into the laptop. He loaded the program and hit the enter key.

He looked at his two friends. "Shit. I've got Oracle in the morning." He went back to iron his dress shirt. The last thing they heard from him was mumbled: "If anyone needs me, I'll be on Ellison's yacht."

Angela studied Naqeeb. "How long do you think?"

He shrugged. "A few days at most he said. "Let's button up the laptop and Xbox."

Angela agreed and added, "And clean up the apartment a little too."

The place was pretty trashed. Pizza boxes and sandwich wrappers littered the kitchen. Empty sodas and energy drink cans adorned every flat surface. Dishes festered in the sink. It took them 45 min to set the place in some semblance of order.

"How did Joe get out of cleanup again?" Angela asked Naqeeb when they were done.

"It's a gift," he said sourly.

"What are you doing tomorrow?" Angela asked him.

"I've got to call IBM and get out of my interview. Then call Google and accept. I need to fill out the offer letter and send that in," he said stretching. "What's on tap for you?"

"I'll start with Stanford's app in the morning, she told him. They kissed.

As they drifted to bed, they heard a muffled voice from Joe's room: "Crap! Cain lost to the Rockies 4-0."

Definitely not the Giants year.

CHAPTER 9

June 1 0922 2010 Monday, Mountain View Ca.

Sitting at her desk Peg Pullman had a frozen sour look on her face. The Irish coffee, half-finished sat forgotten in her hand. She watched CNBC intently. If she tried hard she might be able to will the stock price up. Today had been the bloodbath many had predicted. Stock down 14%. The analyst call yesterday had focused squarely on the Q1 results and Scimitar projections. She'd tried to focus their attention on the "new" product but no dice. The Anderson ouster was all over the news and had provided more bullets for those looking to inflict damage. Even her Gallup poll numbers had taken a hit. They were teetering.

Even the planted question from one of Cantrell's contacts about "Product diversity" couldn't get the bulldogs off Scimitar. God the media was unfair! When Apple even hinted at a new product their stock went up and people fawned. How in the fuck did they manage that? She should have arranged for a Scimitar to get "inadvertently left" in a bar. *Fuck them* she thought.

Cantrell came into the office. He quickly removed the coffee cup from her hand and replaced it with water. "We have to keep up the campaign schedule," he said briskly. "The Primary is tomorrow", he reminded. Hey, we are going to win big here!" Peg grumbled, sullen.

"Upbeat, Positive, right?" He told her. Concentrating on the speech helped bring her around. "Remember the takeaway is that Scimitar set one-day sales records."

That statement conveniently ignored the fact that the sales were only 25% of the projections.

Spin control was everything when the illusion was a reality. Cantrell continued the themes trying to get her focused. "I'm the best person to get the State back on track. I have a proven record with business." Back on Track was their latest slogan. Pullman grudgingly read thru the speech and regained a measure of composure and life. Cantrell went thru her schedule. The event and speech was at Moffett Field. Practically her back yard. The media types loved the Speeches at Ames Research Center. Invariably they used footage of the huge Blimp hangers left over from a different time. The hangers looked retro and futuristic at the same time. Hundreds of car commercials had been filmed in the cavernous interiors. Cantrell looked at Peg, concerned.

"Marcus and George are going to take you over. Eggert has his team in place. "I have a meeting", he reminded her. I'll meet up with you at the fundraiser at the Westin tonight."

Peg barely acknowledged this but it did give her another target for her anger. The Republican Party was trying to squeeze more juice out of Silicon Valley.

She scoffed. "Ungrateful Sons a Bitches." Slurring just a little. "They think we are an unlimited ATM for their cause."

Cantrell let her fume then cut in. "We have to have some kind of party apparatus for the general election. We need those precinct captains and phone banks to get out the vote. The Tea Party gets out a hardcore base for the primary but the party can furnish the swing voters who can deliver the general." He tried to reason with her, thru the haze that the crushing Scimitar numbers and the alcohol was putting in her head. Some of it got thru.

"When do we tack to the center, she asked.

Cantrell nodded in approval. "June 10th he answered. Tomorrow we cruise to victory. We need about a week to mend fences in the party. We stop attacks on gays and illegal aliens and shift to Boxer, Obama, and Washington." He paused. "I'm getting feelers from the Boxer camp about debates."

"Is that what this meeting today is all about?" Peg asked shifting the conversation.

"It certainly is," he lied easily. Now go freshen up, they're waiting.
Knock em dead!"

She went to the bathroom while Cantrell grabbed his briefcase and left the
building. He didn't feel guilty about the lie. This might be nothing.
5 short minutes later Cantrell pulled in the small parking lot off of Middlefield road. He
kept about his business and went inside. He ordered a dozen donut holes and a coffee.
Cantrell took the food back to his car. The man following him got into the passenger seat
without a word. The man took an offered donut hole.

"What's up Larry?" Cantrell asked around a mouthful.

Larry Kocinski looked at Cantrell steadily. "I have a Personal Services
Consulting contract that I want FastFac to sign," he said without preamble.

"And why would we do that?"
Kocinski pulled the iPod out of his suit pocket. He had even shaved for this meeting.
Had to look professorial.

Cantrell put the proffered earbuds in. Kocinski played the audio file.
The background noise was a buzz and then a young man's voice: "The main point is that
the Nanotech is going to be disruptive in ways we can't even think of right now."

Another voice: "I agree. We are on the tip of this thing now. We only know the
barest part of their abilities." The file continued. A female voice joined in while options
were discussed. Finally one of the males said: "Honey, this is big. And dangerous."
Sentient Nanotechnology is a whole can of worms." Cantrell grunted as the file ended.

"Play it again," he commanded. Kocinski complied.

Phrases kept exploding on his brain: Barest part of their abilities. Government
throws a wet blanket on this. Big and dangerous.   No shit, Cantrell thought.

Kocinski put away the iPod.

"How do I know it's real?" Cantrell asked.

"You don't" Kocinski shot back.

"How do *you* know it's real? Cantrell snarled.

"I know those kids, he lied. "They did it", he said quietly. Stanford set them up to do just what they did."

Cantrell knew Stanford had Nano research projects ongoing but his people had told him they had achieved only modest success.

Kosinski confirmed his thoughts. "That's what I thought 36 hours ago." Look, they think they can keep one of the greatest scientific/technological breakthroughs in history under wraps." He ramped up the enthusiasm. "While they figure out what to do, you and I have a window." I can read a stock ticker, he told Cantrell. "You people are desperate. I'm desperate. We can work this out! We have a Window!" Kocinski repeated with a touch of mania.

Cantrell rolled it over in his head. He was intrigued by the recordings. He knew Kocinski from the work they did at Stanford. They had even exchanged hellos at the speech a couple of days ago. He also didn't trust Kocinski for shit. "I want safeguards, he started.

"Tough shit," Kocinski broke in. We don't have time to dick around!"

Kocinski looked Cantrell dead in the eye. "This blows up and you guys have me killed. But trust me when I say, if I end up dead, Pullman dies the next day and you the day after." Kocinski lifted his shirt to show the butt of a gun.

Cantrell goggled. "What the fuck!"

"You're not listening!" Kocinski whined. "This is a straight trade. A ten million dollar secured consulting contract. You got 24 hours to post the bond. I give you the names and addresses." Suddenly you have the single most important piece of tech since the microchip." Kocinski let that sink in. But don't try to get cute he warned I got people on all the sites. One call from me and the feds get it all."

Cantrell chaffed and stalled for time to think. "We're taking all the risks".

"For all the reward" Kocinski oozed. 'Come on! Sentient Nanotech! Each one of those Nano-agents is going to be worth a million dollars. I'm sitting in the car with the world's first trillionaire! Every large tech company in the world is dabbling in Nano. Even you!"

Cantrell knew the Carmel complex was working on those lines.

Kocinski could see the wheels turning greased by avarice. "Jimmy, this works. You're desperate, I'm desperate," he repeated. Those kids are trying to hide it. You walk in and take it what the fuck they gonna say? "Whaa, someone stole what I stole?" Kocinski scoffed. They can't go to the cops or they would be in trouble!" You guys are set and I'm sitting back happy as a ten million dollar clam."

That penetrated for Cantrell. He nodded slowly. "Lemme see this consulting contract."

Kocinski reached into his backpack.

Cantrell halted him. "Larry if you're fucking me on this…

"Yeah, Yeah, Yeah, Kocinski muttered. "Tell me something I don't know. He reached back into the backpack.

Cantrell drove slowly out of the parking lot. He turned onto Moffet Field Road to get to Peg. He had a lot to do. He had to line up the accountant to transfer the money. He had to coordinate Eggert and the Musketeers to start surveillance. Then figure where they were going to do the work.

That Peg would want this was a given. She trusted his instincts on these things. And boy his instincts were screaming at him: Grab it!! Showing his driver's license at the gate, he continued to the pavilion area.

He smiled ruefully. Kocinski had played him pretty well. The 10 million figure was in the sweet spot. A million and they tell him to go take a hike. 50 million and they would figure a way to cheat him. 10 was perfectly positioned to say "I have something" without bleeding FastFac dry.

Oh, Peg would find a way to fuck over Kocinski, Cantrell was certain of that. 10 million was 10 million after all. He'd read in the Journal that a certain Swiss banker was not expected to survive the car crash he was in last week. Cantrell shrugged mentally.

The less he knew about Eggerts and his team's movements the better. He pulled into the Ames research Center parking lot adjacent to the pavilion. Nice crowd. Peg finished her speech to warm applause. Cantrell got Eggert's attention and had Peg brought over to the car. She joined him inside, with Eggert and two of his guys standing guard.

She was fully on board as soon as Cantrell laid it out for her. "Get it done, quickly." was all she said. She left for the limo and the fundraiser. Cantrell got the press liaison over and started rearranging the schedule. The candidate had some pressing matters to attend to and she would need to devote some time to running her business. Wheels started turning.

Joe drove slowly thru the parking lot at the apartment complex. He was looking for a visitor slot without much success. He didn't usually arrive home at 5:30. Monday's sucked. Especially today. Today had been like a real, no shit job. Sighing, he finally gave up and parked on the street. He could have had the spot in front of the apartment but that fucking Comcast van was still there. Someone's cable was hosed, he thought. He locked the car and loosened his tie. He was tired. Tired of smiling. Tired of being nice to strangers. Tired. And he had the morning shift at Peetes tomorrow.

He hung up his backpack as he entered the apartment. Setting the new Kenneth Cole briefcase on the counter, he saw Nange on the couch. Nange was the new couple name for Naqeeb and Angela. He decided that they needed a cool couple name like Brangelina or Bennifer. Nange had barely grunted when he came in. Joe could see he was second fiddle to a new life form. Well, duh! He thought. Still. They continued typing.

"How was my interview with Oracle, you ask?" Amazing, thank you!" The chef prepared bacon wrapped bay scallops for lunch and they served it with a dry pinot that was excellent, thanks for asking!" Joe looked at them. Nothing. Fucking ingrates! He huffed sitting in the chair.

Naqeeb grinned. "Dude! Messin with you!" Spill it."
Angela laughed.

Joe had to smile. He might have deserved that one. "They want to use it to manage the cloud, he told them without preamble. "That's something I hadn't thought of, he confessed. They think you can organize data into predictable segments. Then it's easier to store and retrieve."

Naqeeb was nodding. "Kind of like defragmenting your hard drive back in the DOS days."

"Yep."

"Cool."

Angela asked the practical question. "Did they make you an offer?"

"One very similar to Naqeeb's from Google, he told them a matter of fact.

"Jackpot, Dude!"

Joe was silent to the general congratulations.

"You're not still thinking of... Angela started.

"New Star Technologies, Joe cut in. "That's the name of the company and yes I am." "I'm going to take their offer like I told you."

"Not cause of this, Naqeeb gestured at the Xbox.

"I told you before that I was leaning towards New Star and nothing Oracle said or offered changed my mind." Joe had the whole thing settled in his head. "Now what's going on with System? He asked.

"They are trying to make sense of Internet protocols and programming languages," Naqeeb told him. "Can you make sense of this, he gestured to the screen. The response from System was a weird hybrid of code and English and different statements.

Joe studied it. "It looks like code but with different things thrown in, he said. Weird Windows DOS stuff and dBase it looks like," he said.

"It's in octol, Naqeeb told him.

"We need it in binary or even decimal to compile it, Joe complained.

"Can you tell System about the different languages and binary to machine code? He asked them.

"That's what we were doing, Naqeeb said pointing to the screen. "They have to learn a lot of languages to absorb your program. But first, they need the basics."

"Are they finished with my program already? Joe asked surprised.

"Not yet, Angela replied. "See, they are just now looking at the internet in response to the data mining portion."

Joe looked at her surprised at the computer lingo from her. Angela knew about computers, just not the nuts and bolts of programming. "Well, that's what Naqeeb says, she added defensively to the look from Joe.

Joe smiled. "So really, they are looking at interfaces and protocols. He got up from the chair and got a redbull from the fridge. Think about that, Naqeeb. They have to go back and learn DOS, then Windows, Internet Explorer, The IT/TP protocols, all the network software from Juniper and all the webs, HTML stuff. Naqeeb whistled.

"And my program is in JavaScript with HTML/canvas for the data mapping. The predictive matrix is C++ with tables linked in dBase."

Joe looked at Angela who looked confused. He smiled. "Ha, now you know how I feel when you talk most of the time!"
Angela changed the subject to regain the upper hand. "We have to show the CD to Joseph," she turned to Naqeeb and pointed. Her boyfriend grunted and popped the CD out of his laptop.

Joe used the handheld microscope to see the fine web of lines covering the CD. He glanced at Nange. "They are thru with this?"

"Yep, said Naqeeb. "They compared what was on the disc to what was stored directly on the hard drive, he informed Joe.

"Did they ask before moving onto the CD?" Joe asked.

"Yes they did," Naqeeb answered with a smile.

"Good." "I do have other copies of the program so this CD can be quarantined," he told his friends.

Naqeeb got up from the couch and stretched. "I have a couple of thoughts along those lines," he said looking at the CD. "We need to see how tough these guys are."

"Define tough," Joe said.

"We need to see what it takes to destroy them, Naqeeb said flatly. That set-off Angela.

"We can't kill them, she cried. They are alive!" "We've just reached a point where we are communicating better, understanding them, and now you want to kill them?"

Naqeeb looked at her steadily. "We agreed without controls, System could be dangerous. We must know what it takes to destroy them and what environments they can survive." His reasoning caused general silence in the room. 'System told us that the agents on the CD had been sacrificed for the understanding of the program, he reminded Angela. "They're dead to System already."

Angela refused to acknowledge this fact. She looked away. Naqeeb appealed silently to Joe.

"Dude, I agree with you in principle but I'm going to back Angela on this. She's the mom here." Joe looked between the two. "If she says yes, I'll go along, but I think we need to be real careful before killing off parts of System", he concluded.

The man thought that over. "My point is that we don't know what kills them. They've only been exposed to water air and vacuum. What if direct sunlight kills them? Or screws them up? I'm not saying it will, but a testing program could keep us from inadvertently killing many more of them." He moved around the kitchen and living room throwing out questions. "Think of what we don't know. Can I run my finger along these lines on the CD without dislodging them? "Can I drop the CD without them coming off? Can I crush them? What about a flame?" "High heat?" he finished the stream of consciousness.

"Are you suggesting burning the CD?" Angela asked him scornfully.

"No," he said gently. But we used the handheld microscope to look at them in the Xbox. Did we kill thousands when we did that?" He waited her out.

"What kinds of tests are you proposing?" she finally said.

"Honey, you are the materials engineer. Honestly, my first thought was to drop the CD into a bath of water but then I remembered the fluid medium was water, Duh!" Smiles greeted this.

'The water medium was kept at a pretty narrow temp range," she reminded her boyfriend.

"Exactly"

"Maybe they are like us. Drop us in 105-degree water for long enough and we die right?"

"Same with cold," Naqeeb reasoned back at her.

"At some point, we are going to ask System to go into a dormant stage and we need to know the ideal storage environment for them." "Temperature, humidity, pressure, all those factors," Angela said.

Joe could see she was coming around. The engineer in her warring with her motherly instincts. Interesting. Nange sat back on the couch rattling on about tests. A hammer? Sheesh!

"So System is learning the internet and my program for the next few days and they are finishing up the new worker bots right?" Joe broke in on the test designing.

'And you and Ms. Frankenstien are going to be experimenting on System for the next few days' right."

Naqeeb nodded happily. "Yep, why?"

"Don't you have to work at the Kitchen?" Joe asked.

"I told Google I'd start on the 28th, Naqeeb informed him.

Joe nodded. "Angela?" he asked.

"Stanford has it all under control. They told me today that the grant money should be in hand on 4 Oct, provided the government cooperates." They want me next month to start some busy work that they are going to take out of hiding," she told Joe loftily.

"That starts a new fiscal year, he told her. Better get used to dealing with all that government funding crap."

"Morgan Hill, huh? That's a nice area. You two should be happy there. Close but not too close…" He added slyly.

Angela shoot a look at Naqeeb at Joe's non sequiter. "Did you tell him?"

"I did not," he told her defensively. "You've been with me all day! They both looked at Joe. "Dude if you ran us thru your little rat maze…"

"Relax, Joe cackled. "I saw Angela's browser history. Morgan Hill apartments. I could, of course, tell you exactly which one..."

"No thanks!"

"Hey, suit yourself."

Joe sauntered back to his room. 'Got to run, going to see Renee before she goes back home."

"Dude, Naqeeb called, when you gone?"
The muffled voice floated back. "Not quite sure. "Giving my notice at Peete's tomorrow. Pretty soon after that, I'd guess."

Peete's coffee was dead the morning of the second. School was out and a new Starbucks closer to the medical center was killing them. That and the lousy service, Joe thought leaning against the counter lazily. The hipster with the soul patch and the fedora ordered a half caf/half decaf soy mocha latte in a small cup with extra foam. Joe rang him up and wished he'd lose the hat and facial hair. Tired and played.

"Large coffee, please" said a female voice.

Joe kept his face passive as he rang her up. "Thanks, I like those simple orders", he confided. "Room for cream?"

"No thanks."

"We're just brewing some fresh if you don't mind. It might take a minute," he told her.

"I've got time."

"I'll write Kerry on the cup to remind me of you since you look so much like Kerry Washington." The woman smiled.

"Kind of slow in here," she said.

Joe busied himself getting the hipster out of the store. "Yeah, schools out and the summer session hasn't started yet."

"Do you go to school here," she asked.

"Yeah, just graduated," he told her.

"Wow, must be exciting. Stanford grad. Fielding a ton of offers?"

"One good one I'd like to take. Tech Company out of DC. Kinda cool."

"Moving out of sunny California to battle the snow in DC", she teased.

"That's going to be a problem, he admitted. But I got to be out of my apartment by the 30th. I was hoping to start the 28th if they offer it. I need to swing by and say bye to my mom."

Joe poured the coffee in her to go cup and handed it over.

"Sounds like you got it all figured out," the woman said. Then a mumbled "pain in the ass".

"I hope so. Have a nice day!" With his biggest smile. She left the shop.

Joe nodded to himself. "Hey, Mark! Do you have a minute?" he asked the day manager.

Approaching the van cautiously from the blind spot, Tyler Eggert banged on the back door. His training would never allow him to approach from the front of a car. Even his own. Training and duty were all. If Cantrell and Pullman wanted him here then here he would be. I'll get rid of the amateurs first. He pounded on the van quarter panel. The van door opened and a man in an orange vest, hard hat and jeans stepped out. "We're working the cable outage here", he started.

"Stop. Leave. Now." Said Eggert.

The man stepped back at the look on Eggert's face. He began again, "We are authorized to work…"

"NO" Eggert overrode him and looked at him.

The hard hat finally caught on. He shrugged. Alan wasn't paying him enough to mess with this guy. He closed the back door and went around to the front. He unlocked the door and slid in starting up the van. He left as quickly as he could. Allen had said the new team would want a turnover of what he knew of the situation. That guy was too grumpy to want an exchange of information. His loss. He passed a dark SUV coming in as he went out.

Eggert met Porthos and Artimus at the vacated spot. He always thought of operational names. He got into the front side passenger seat. He sketched out the operation. "Apartment 210E. Three occupants- 2 males and one female." I want full movements and tabs on what they are doing in there."

Eggert turned to face the black, ex-navy man Porthos. He was the computer expert. Pay attention to computers and anything to do with nanotech. Do you have any background in that?"

Porthos shook his head.

Eggert reached into his suit jacket inner pocket and removed a folded sheaf of papers. "This is all we know about the subjects." He handed over the file to Artimus in the driver's seat. "I want a full work up in 24 hours.

The woman scanned it and handed it over to Porthos. He started fiddling with the surveillance gear in the back seat. She asked Eggert about resources.

"Dartagnion and Athos will be here in 47 min. They will have long term cover."

"I hope it's not another Comcast van," Porthos joked.

Eggert did not smile. "Get this. I have operational control on this op. Do not suddenly decide to do something. You watch and report. We go in on my signal. Got it!"

Artimus responded in the ubiquitous army: Oh rah!

Naqeeb glanced out the living room window as the 2 SUV's drove off. The Comcast van that had been there forever was gone! He was tempted to move his car into the spot.

It was still on the street but at least he could keep an eye on it from the apartment.

Angela called him back to the kitchen. "Water's boiling at 211 degrees she informed him.

He returned and took up the piece of the CD they were testing. "Ready with the stopwatch"? He asked.

"Affirmative. Ready, go!"

He dropped in the piece of CD.

They were systematically killing off the bots. The agents were remarkably tough but they could be destroyed. Acid, for instance, left no trace. But they could survive a fall. Simply running your finger across the lines of bots would not rub them off. You had to scrape them off with an object. They did have a limited ability to repair a break. A thin knife cut had been healed but the one-inch gap could not be overcome. Angela figured they didn't have the resources to bridge that far. Sunlight and magnetic fields seemed to have no effect, but they couldn't be sure without further tests. If only they could subject them to high energy blasts, she could... Angela mentally shrugged. Another item for her notes.

"Time."

Angela retrieved the CD from the pot. Clean as a whistle. Uh-oh.

They both looked at the pot of water. Shit! Now they had to boil down the water to see if the agents were truly destroyed or just suspended in the liquid. Angela sighed. They just didn't have the controls the tests demanded. Naqeeb had been right about needing to determine their limitations, but this way was primitive. Besides, he didn't need to know he was right about something. That set a dangerous precedent.

It took almost an hour for the water to boil off. Naqeeb scooped up the whitish residue and looked at it using the microscope. Nothing.

"The heat and agitation of the water seem to break down the bonds, Angela theorized.

'What about cold?" Naqeeb asked. "I suspect that extremes say 30-40 below zero would render them inoperative, Angela answered, but we need better facilities to do that kind of test."

Joe entered the apartment as she said this. He looked between the two. "Trouble?" he asked.

"No more than usual", Angela told him.

"Cool. How are the kids?"

Naqeeb showed him the results sheet he was keeping.

Joe read thru it. "So kind of like us coupled with sensitive computer equipment", he said.

"Yeah, that's what we think too", Naqeeb confirmed.

"And we have two more days before System is through…" he started.

"I think they will be done tomorrow with both the worker bots and your program, Angela said.

Joe absorbed this. "So they would be free to explore", he mused to no one.

"I heard from the DC company today", he told his friends. "I should have a firm offer and a start date soon. I gave my notice at Peete's. I'm all done." he said with finality. Nange expressed their happiness.

"Yeah, so I'm out of here on the 10th or 11th. I want to swing by Hemet and see my mom, Joe said. "Then I think to take the 15 to Vegas, before hooking up on the 70 and taking that across."

Angela was silent while Joe laid out his travel plans. It hit her hard. That was funny. She was annoyed by realizing she would really miss Joe. He was Naqeeb's friend and roommate, but now she was sad to think she wouldn't see him on a daily basis.

"Dude, you should go on Couch surf.com to find places to stay while you travel, Naqeeb advised.

Joe nodded. "Good idea. So that puts me in DC around the 21st with a week or so to kick it before I start." "What do you guys want to do with the stuff in the apartment?' he asked them.

They all looked at the valuable, Salvation Army furniture. Joe wandered around looking at their stuff. "I figured maybe you guys could take it over to your new place and then replace it slowly with decent stuff?" "I'm going to rent a room near work, to begin with, so I don't need anything."

Naqeeb nodded. He liked that plan. He looked at Angela who had been silent. "That okay honey?" She agreed. Angela looked at Joe and asked if he wanted anything from the place.

"Well, I do get the Xbox! You promised!" He told Naqeeb.

The friends all laughed.

"Seriously, I don't need anything but my computer and TV. You should take my bed and put it in your second bedroom. That way Naqeeb has some place to stay when he's in the doghouse."

Angela smiled. "More like a bed for Khush when he comes to visit", she said. Naqeeb grimaced. 'No way that kid hides out at our place. He's on his own!"

They spent the evening dividing up the apartment's treasures. Joe went out at 10 to see Renee. "I can't do this by text," he told them. "I swear I will clean the bathroom and my bedroom before I go". Nange was skeptical. He slid out before they could pin him down. As he walked to his car, Joe cursed to see the PG&E van parked in the spot across from the apartment. First Comcast, now PG&E. Corporate America was trying to get his car stolen before he had a long trip to take. A holes.

CHAPTER 10

June 2 0745 2010 Tuesday, Palo Alto, Ca.

Again, Naqeeb looked out the living room window. The morning sun tried to break thru the marine layer. Might be a little warm if the fog didn't hold out till 10 or so. He noted the absence of the PG&E van. They had finally fixed whatever was going on. He hoped the new place in Morgan Hill had better parking. He looked back at the apartment. The place was a mess. Boxes and suitcases and food containers warred with computers and notebooks.

Joe and Angela were on the couch ready to start with System again. The three roommates were living their lives in weird thirds now. The first third was their old student life. Harold had indeed just "dropped by" to see what was up on Monday. They just shut the laptop down and flipped on the TV.

"Nothing much, what's up with you?"

He couldn't pin them down.

The second phase was the new adult side of life. Stanford, Google and New Star all wanted a million forms filled out and drug tests and information. A credit check to rent an apartment? The cat and the snake had to have new homes found for them. Suelyn took in catpuchino, while Estaban the snake went to Mike. Yeah, he looked like a good snake daddy.

The third proved to be the weirdest. Creator to a new life form. System was proving to be fascinating and exasperating by turns. By Monday noon, they had indeed finished the worker bots and the understanding of Joe's program and were happily exploring the internet. Was Joe pleased? Angry? When System improved his program. Upset he may be, he still commanded System to copy the improvements to a CD so he could look them over. It took him hours to understand what System had done. Mind boggling. "Dude, he told Naqeeb. Look at your standard password hack program. Most of the time you approach it as to what you know of the human: birthdate, kids names that kind of thing. Then you just try different combos till it cracks. The other way is to figure out the computer system: how many characters, and what is acceptable and then same thing- just try different combos until you hit the right one. That's how most detection software gets you. It sees you trying millions of different combos and shuts you down."

He paused making sure Angela was keeping up with the programming talk. "System looks at the other computer first using low-level stuff to cut down on the number of false tries. Then integrates what they know of the user to come up with the most likely password. Much more efficient. Same thing as my program- System integrated the machine and the human in the data mining and the predictive matrix. Makes it much more accurate and efficient", he grumbled.

Angela thought this over. "The integration is really just who they are, Joe. A blend of machine and organism."

All three sat back and thought about that. To really understand computers and people on both levels was difficult at best.

Naqeeb reviewed the code with Joe. "System can really get the computer programs to spill their guts under the guise of low-level protocol transfers, he asked?

Joe pointed. "See this is where the OS admits to the registers in the field and here is the register to valid character correlations."

Angela let them discuss it for 5 minutes or so. "Can you dumb it down for me, please?"

Joe smiled. "I'm going to remember this moment." "Look at it this way: when two people meet they shake hands and one person says "How's your day going?" And the other person says fine, how's yours? Then you start in on the real conversation."

He paused getting up from the couch and pacing. "With computers it's similar. One comes on and says "I'm here and here's who I am and how I talk. And then another responds. Then they exchange information or not based on who wants what." Okay so far?"

Angela nodded.

"But sometimes when a human meets the right person who has a sympathetic ear and strikes you right you just open up and start talking to them about deep shit right?"

Angela again nodded. "Yeah maybe, sometimes."

Joe wrapped up. "So System just gives the computer a warm firm handshake and says, tell me how your day is going, I want to know all about you." Kind of like the computer equivalent to the local bartender." Joe finished pacing around and looked at Nange.

Angela drew breath but Naqeeb interrupted. "But really important stuff is only on trusted networks or sites." Angela agreed with the thought.

"System is able to mimic trusted sites and networks on low-level items over and over again to build trust. Once it has a trusted network they use that to prove to other sites they can be trusted. Kind of like having a female wingman," Joe concluded.

"Some data is air gapped, Naqeeb argued. And, there are stand-alone networks that no one can get to"

Joe thought that over. "That's tricky. That's how most DOD and Intel sites are handled.
I wonder how System would handle that."

"Don't give them any ideas" Angela told him.

Joe agreed. "Besides, everything is digital these days. All of that data is stored somewhere. Electronically. It's just a matter of finding it and getting at it, he finished dismissively.

That line of questioning led them to some of the softer spots: Does System feel? Do they have emotions? Anger? Love? Loyalty? The answers from System were frustrating. Mostly they responded with "Unknown data" to a question they didn't understand. System could differentiate between billions of people but could not understand "emotions".

The trio spent hours explaining human emotions and motivations to System. With varying degrees of success. However, from these sessions, truths emerged. System would honor the agreement they made with the creators. System did feel some loyalty to them. The rest of humanity? Not so much. The protocols were viewed as commandments. The agents did understand the concept of death and would not kill or sacrifice parts of itself without great need. The nanobots had a great deal of information but were incredibly naive about using it. System would tell the truth in almost all instances but was unaware when that truth might be embarrassing for another. For instance, System had no problem revealing that Joe had $123.45 in his checking account after he paid for a pizza, Sunday night.

"123.45 Joseph?" Angel scolded, "You never save."

"Hey, that's just my checking account. I got a couple thousand in savings," Joe said defensively.

System had no frame of reference for money or wealth. "Resources should be apportioned according to needs and priorities," they replied to an inquiry.

"I had no idea you were raising the kids as Communists," Naqeeb chided.

They powered System down (actually the laptop) at 8:00 pm per normal. Naqeeb theorized that they needed the sleep just like humans. "Humans use sleep to align emotions thoughts and information obtained during the day", he said.

"Thanks, Dr. Freud", Joe cracked.

So Tuesday morning saw the three in a complete state of flux. Naqeeb turned from staring out of the window to the couch. Joe was lobbying hard to test out Systems data gathering capabilities.

"We only have a couple more days before we have to put this on hold for a little while. Let's see what these guys can really do!"

Angela was more cautious. "Nothing classified or involving money", she warned.

"I have something a little more altruistic in mind," Joe said.

"Spell it out, dude."

"Well, you know I have a pretty good map of the BART System, right? " For my thesis? He reminded them. I specifically looked at train times and individual disruptions to the on-time performance." Nange nodded along with him.

"I was wondering if System could look at BART on a macro level and try to improve the whole system. Make it more efficient. More bang for the buck, so to speak." Joe looked back and forth.

"Could System *Improve* BART?" Naqeeb wondered.

"There have to be parameters," Angela said.

"Exactly, Joe said enthused. "We can't just have System say: throw piles of money at BART. One of the limitations has to be revenue and fare increases along already approved lines."

"Yeah, the legislation is not just going to give BART a big pile of cash. What would System recommend given the constraints?" Naqeeb got excited about what Joe was thinking.

At 8:00 am "Hello creators," was displayed on the laptop.

Angela typed: "Good morning System. The creators' request that you *understand* and map the Bay Area Rapid Transit System. List all recommendations to optimize BART performance.

Joe jumped in. "Tell them that no further resources other than planned fare increases, capitol outlays, and planned improvements can be taken into account." Angela typed furiously.

"That's good Naqeeb said. Have them list all equipment, software and personnel upgrades as necessary."

Angela mumbled as she typed, "I'll have them estimate how long to complete the task."

"System will comply, 5 hours to complete task" was the reply.

"Good I've got more packing to do", said Joe.

"Hey, don't forget to vote," Angela reminded him.

Joe grunted something that might have been "yes" or "no", and drifted back to his bedroom.

Pullman and Cantrell sat tensely in her office. The primary election all but forgotten in light of the new opportunity being offered them. Eggert was shown in. He sat in front of the desk, his basic gray suit neat, and clean. Clean shaven face all planes and angles and unreadable to Peg. He certainly looked like an ex-military person, she thought. He considered his notes and started.

"It appears they are sitting on something, he said quietly. "The information from the source, he refrained from naming Kocinski, was correct."

Both Peg and Cantrell started breathing again.

"Three occupants in the apartment: Angela Chin, Naqeeb Al-Meri, and Joseph Smithson, he continued. All Ph.D. graduates from Stanford. In fact, you spoke at their graduation ceremony on the 29th."

Cantrell shoots a tight look at Peg, who smiled. She appreciated the circular nature of things. She had spoken of innovation and hinted at a new product and now here it was. The fact that she was going to steal to innovate never crossed her mind.

Eggert continued forward. Always forward. "The technology is contained on a laptop on the premises. The Nano-agents have apparently infected the laptop. They have spent the last few days trying to figure out how to destroy them." He looked down at his notes. The principle is the female, Angela Chin. She is the Materials Engineer. Smithson and Al-Meri are supporting players."

Cantrell cut in quickly. "Can we get it?"

Eggert stared at him thrown off course in his litany. "We have a window," he said, unknowingly echoing Kocinski. "The Nano agents belong to Stanford but the students stole them. They all have job offers and are packing to leave the place. I pulled my team out last night. Even civilians will notice if someone hangs around too long." "We can get the laptop", he concluded.

They discussed logistics and timing. The head of Security assured them that a two-man team could get the job done. "A simple smash and grab. 2 min tops inside the place."

Cantrell nodded beaming at Peg. "I told you! Who are they going to tell? The police?" The first question the cops are going to ask "is what was on the laptop." Very low priority for them." He got up from the chair and went to the windows, then turned back to Peg. "Stanford can't really even go to anyone. They look like the keystone cops, asleep at the wheel while children stole valuable research."

Peg pulled him back to practical matters. "What do we do with the laptop once we have it?"

"Eggert's team delivers it straight to Cooper at the Big Sur complex. He's been told a few things and is ready to go." Cantrell started back to the desk. "He-"and stopped. He looked at Eggert. "Tyler, thanks for the hard work. Go get your people ready for the campaign events tonight and the side job. We'll tell you when."

He held up a hand ushering the security chief out. Pullman stared at the man's back as he left.

"Is he going to be a problem? She asked.

Cantrell considered. Eggert was in line for a substantial pay raise. He knew where all the bodies were buried up to a point. However, his hands were not clean either.

"I'll handle him. He has just as much to lose as we do", he reminded her.

"How long will Cooper need,"

Cantrell frowned. "He's non-committal. Anywhere from two weeks to two months to run tests and figure out general characteristics and capabilities. We can leak it then when we have some data, Cantrell told her. He breathed in and exhaled the cold clean air of greed. "Christ the speculators will drive up the price at least 10 times, maybe 50", he said. "How does a thousand bucks a share sound to you?"

"If the agents are even half of what we think you'll be the richest person in the world. The notoriety alone might make you a Senator," he concluded. Cantrell paused and threw in a note of caution. "The Feds still might interfere".

"Fuck them, Pullman said coldly. "Should I say anything tonight?"

It was a mark of their excitement that the primary results were of minor concern next to this project.

He talked her through the strategy. "Maybe a hint. "New and exciting days for yourself, the company, California and America are ahead!" He reviewed his notes. "Polls close at 7:00. The majority of returns should be in by 8:00. I expect KRON 4 to call it around 9:00 pm. You go on at the Shoreline around 9:45."

Peg giggled gleefully. "Right in Google's back yard!" "That's perfect". "What about the rest of today? She asked.

Cantrell went through it again patiently. She had a tendency to forget details in the manic phase. "You go for the voting photo op around 10 am. That makes the noon news, he said. Then we lay low until tonight." Internal polling had them leading by 12 points. "Tomorrow is a normal day at the office. A round of calls pitching for unity in the Republican Party to defeat the Democrats," Okay?"

He didn't need to tell her that tomorrow would be the op for Eggert's team. Peg's mind whirled with the possibilities. She motioned Cantrell out. "Go take care of this, She told him.

Cantrell saw Tyler Eggert in the waiting area outside Peg's office as he left. He motioned him into his own office down the hall. "Tomorrow, he told the head of Security. Just the laptop right?"

"It contains her notes and the agents," he confirmed.

"That's good work, Cantrell told the man. You are in line for a bonus."

"That's always appreciated," Eggert said.

Cantrell handed over a form. "Drop that by accounting. A thousand options at today's price." Six months from now that paper could be worth two million dollars."

Eggert said nothing, just smiled and pocketed the form. Cantrell sat back pursing his lips. "You know what, he said. I just thought of something. Maybe Kocinski takes that 10 million and buys stock with it. 500 million potentially." "Kocinski could be sitting pretty, he concluded. Huh."

Eggert cleared his throat. "Speaking of Kocinski…
        "Not now, Cantrell told him. Plenty of time for him. Let's talk about tonight's events." The two men got down to business.

The roommates got back from voting after lunch. "System Ready" was displayed on the Laptop. Joe whooped. "Let's see what they got!"

Angela typed on the computer asking System to display the results. "Have them put it on a CD as well," Naqeeb asked her. She complied. System complied to that as well. Naqeeb's laptop whirred while they read thru the recommendations.
        Joe brought up the recommendations on his machine when the CD was created so they all could look at it without crowding around the laptop screen. He was getting tired of scrunching down to pear into that screen.
        The three finished a quick reading about 10 min later. Astonishing! The list of recommended improvements was 17 pages long.
        "Basically, blow it up and start over", Naqeeb said reading thru.
        "No, not quite, Joe countered. "Look at this," he pointed. BART has some antiquated ways of doing things. They still have a manual system to monitor tracks and train movements. Manual because the union will not give up the jobs." Joe told the pair. "System wants to automate that process but it recommends moving those jobs to station agent jobs." BART would have to negotiate that with the union but if there is a net zero loss of jobs, the union might go for that."

Naqeeb and Angela agreed that stood an even chance of getting through the contract talks.

Naqeeb was scanning further down and looking at files on Joe's CD. "Joe look at this bit. System rewrote the interface between BART's maintenance and logistics programs."

Joe whistled softly. "Those are hella old. We're talking 1980's code. I know that would be prohibitively expensive to replace both programs at once."

Angela smiled at Joe. He was like a little kid with a train when it came to BART.

It took them the rest of the day to digest what System had provided. The sum was a plan to make BART better for a little less money. Joe was bouncing off the walls. "I know you guys had reservations about my software, but look at what they did! This is the kind of thing we could do with these guys!" "We have to do something with this", he pleaded.

Nange agreed with him but how to proceed? Angela was silent a minute before she spoke up. "My dad knows a guy...she started.

Joe cut her off grinning. "That's an excellent mob phrase: "Hey! My dad knows a guy!" His Brooklyn wise guy accent was terrible.

"Joseph!"

"Sorry"

"My dad knows a San Francisco City Supervisor, Angela continued. "Ed Lee. My dad contributed to his campaign."

Naqeeb looked at his girlfriend anew. "Whoa, that's kind of heavy."

Angela shot him a look. "This isn't about my parent's money. It's about getting this in the hands of someone who could do something with it."

Joe raised his hands. "Fuckin A right!"

Ed Lee was the director of public works, Angela told them. He is also a politician. He could ram enough of this through to make a difference." She finished up breathing a little heavier.

"And if this puts him in the catbird seat for Mayor?" Naqeeb asked.

Joe answered up. "So what? I don't care! It's win-win. BART gets better for all and he gets to be Mayor."

Naqeeb nodded. "I agree. How do we get it to him?"

'We give it to Angela's dad to get it to the Supervisor," Joe answered.

"How do we know he will do anything with it?" Naqeeb asked playing devil's advocate.

Angela answered steadily, "We don't." "But at least we did something!"

The three started to work organizing the material into a workable plan and crafted a cover letter for Angela's father. System shut down while they worked. The trio watched TV and worked late. The Giants were taking apart Cincinnati. Matt Cain threw a complete game and the new call up catcher Buster Posey looked like a natural. 35-22 at this point. The team was shifting into high gear!

As Angela dropped the CD into a manila envelope, they shifted to the 10 o'clock news. Peg Pullman was delivering her acceptance speech. She finished with 52% of the primary vote in a bit of a shocker. The press lauded the power of the Tea Party.

"There's your girl," Joe chided Angela.

"You took a picture with her," Naqeeb reminded him.

"Yeah, yeah, Joe mumbled. "You guys got the script," he said changing topics quickly.

Naqeeb nodded. "A friend at Stanford did this as a project. Stanford wouldn't put their name on it because of the politics involved. We thought Ed Lee could make some of it work. Could he get it into his hands?"

Joe smiled. "Stay simple and let Naqeeb run it out, okay?" asking Angela. She agreed. "He never thinks I can accomplish anything."

"You'll always be his little girl, Naqeeb told her. He has problems thinking of you as an adult. All our parents do", he motioned to the group.

They made plans for tomorrow. "I got to run a load of crap over to Kelly. You guys hit up your dad and we meet back here around noon?" Cool?" Joe asked. Agreed.

The next day, Eggert emerged from Pullman's office and motioned to Artimus and Porthos. The three went down a level to Eggert's small office. All of his people were tired but this op was a milk run.

Artimus stared at him. Silent, accepting. She was in a lot of ways the steadiest member of his team. A 31-year-old Latino and ex-Navy. No family, no ties, no conscious. She was tall and firmly built with close-cropped hair.

Porthos was another story. They needed the tech savvy but it came in a package of a young black man who didn't like to follow orders. Black and only 28 he was wiry and fast.

Eggert sat them down in his office. "This op should be simple. That's why it scares me", he said. He laid out the plan. "Steal a car from the mall area. Non-descript. Drive to the apartment. Wait for it to be vacated. Go in and grab her laptop. The last data we had shown them packing to leave. You may have to hunt for it. But computers are the last things to be packed by kids. Find it and go. Any trouble and you leave."

He waited for Porthos to acknowledge him. "Drive the laptop to Dartangion, at point Bravo. Dump the car and return here. Got it?" Affirmative from both operatives. "Let's talk contingencies.

Artimus asked some good questions about the roommates. "They are in and out in all combinations all day from what I've seen." Eggert agreed. If they hold up or the apartment is never vacated, we go in at 3:00 am. Stealth."

The pair nodded. Simple plans were always the best. The action team left. Eggert didn't need to tell them to change out of business attire and into street gear. He didn't think the three targets would call the police, but if they did, he wanted the stereotypes playing in his favor. Black man, Latino woman? Dressed like thugs? Give them what they expect and they'll never look deeper. D'artagnion came into his office and sat without a word. "Be at point Bravo at 1000 with the car full and your bladder empty," Eggert directed. "Call me when they deliver the package." "I'll give you point Charlie then."

"Roger that" was Dartagnion's only reply.

Eggert looked at his hired killer. "This needs to go smooth." Dartagnion left his office silently.

Eggert didn't think this would call for killing. Not yet anyhow. Pullman and Cantrell were lunging at this with both hands. Something was up for sure. The 20K was just a start. Once they had the Nano agents, the stock should shoot up. Eggert intended to retire someplace nice and quiet after the election. He knew a few places where 10 million bought you security and anonymity. He was pretty well done with following rich people around. Time to be the rich guy himself.

It seemed to take forever to get the car. Porthos was fuming. Eyes on the apartment didn't happen until after eleven. The omnidirectional mic was picking up no interior sounds. Nothing. And hadn't for the last 20 min. The place seemed empty. They waited another few minutes, with Porthos trying to control himself. Artimus had the lead on this. He looked to her for orders.

"Wait in the car," she told him. "I'm going to knock on the front door. If I don't return in 3 min follow me in." He rogered up.

She left the car. The day was sunny so no hoodie zipped up to conceal her face. The sunglasses helped though. She strode up to 210E without looking around. She started to work.

Porthos fidgeted in the car. He hated this part. The bitch had no idea what she was doing. At 90 seconds he got out and followed the same path as his partner. He slipped on the sunglasses and the Giants cap. The black sweat pants and Nikes completed his look. He continued up to the apartment building with a few furtive glances around. Deserted. He completely missed the red Lexus pulling into the parking lot.

Artimus was kneeling down picking the lock when he joined her. The staircase and landing provided cover from the street. *Three minutes my ass* she thought as she continued on the lock. It gave with a click. Thank god. That idiot was ready to kick down the door.

They entered the apartment and relocked the door. The place was a mess. Boxes and papers littered the kitchen and living room area. There were no laptops on the kitchen table. Shit! They didn't have time to go thru all these boxes.

"Check the main bedroom," she told Porthos.

He left quickly. She started systematically going thru boxes in the kitchen. The clock in her head got to 97 when the key rattling in the door lock alerted her.

Things seemed to happen very, very quickly after that. Artimus slid over to the left side of the door with her back to the outside wall. She would have one or two moments of surprise as someone came in but not much. *Move quickly* was all she thought as the door started to swing.

Porthos emerged from the bedroom at that point carry a laptop clutched in one hand. Power cords and converter held in the other. His face, visible to Artimus was frozen in surprise as he came up to the man and woman as they entered the small apartment.

Shock caused the pair of occupants to freeze and go silent for a critical instant.

Artimus stepped forward calmly from her spot on the wall and took two steps to deliver a punch to the tall man behind his right ear, as hard as she could.

The surgical gloves prevented her skin from slicing open but not from breaking her knuckle. Naqeeb dropped like a stone. Angela drew breath to scream. The other operative continued walking forward, shifting the laptop to his left hand, and gathering in the cords as well. He delivered an open-handed slap to her face.

The scream became a small grunt of pain as she went weak in the knees and her vision exploded into stars.

Porthos barely broke stride and continued to the door. Artimus pivoted and followed him out the door. They both walked to the car unhurriedly, got in and drove away. The clock in her head stopped at 163.

Neither spoke. Porthos concentrated on driving the car to the designated point. Artimus slipped the laptop and power converter into a black carrying case. Hats, sunglasses, gloves all went into a plastic bag.

The man spoke. "It's hers. I'm sure. It was on the floor next to a box marked "Angela's bedroom."

She grunted. At least one thing went right, anyway.

"You shut up and let me talk to Eggert." He shut up.

They met D'artagnion at a gas station near the 101 in Sunnyvale. The black case went into the trunk. The plastic bag into the dumpster. He handed her some lint free cloths out of the window of his car while she leaned in. They would be used to wipe the car down when they dumped it.

"Any problems, you're later than I thought?"

Artimus just looked at him.

"What happened?"

"They came back right in the middle of it."

The special ops vet looked up at Artimus from the driver's seat of his car. "You didn't have a look out posted?"

"We did," she said defensively. "Bad timing."

The senior agent took out his phone. Artimus took on a slightly begging tone.

"Tell Eggert we are coming back as soon as we dump the car."

He nodded and thought: *sucks for you.* Someone has a mess to clean up.

CHAPTER 11

June 3 12:15 pm 2010 Wednesday Palo Alto, Ca.

The ice pack helped with his head. Naqeeb shifted a little on the couch. Angela sniffed and wiped more tears and tried to help. He looked at her thru the fog in his head.

The side of her face was red and puffy. *She might have a shiner tomorrow,* he thought dully. Rage built up again and he tried to rise. The dizziness immediately hit him and Angela laid across his chest telling him to keep still.

The sound of keys in the door caused her to stand fearfully. Joe entered with a smile that changed the instant he saw her. The half-formed "how'd it go?" morphed to

"Holy shit!"

Angela could only start crying again as Joe hugged her.

"Let me see," he said gently pulling back after a moment.

His face was a watery blur of concern as he probed her cheek.

"Hurts?" he asked.

"A little," she said getting herself under control.

Naqeeb groaned and tried to speak. Joe released Angela and went to his friend. The lump on his head was cool from the ice and Joe thought he might be concussed.

Naqeeb could only manage "We, we... before another wave took over.

Joe turned to Angela who hitched out, "we got robbed!"

A stunned expression was on his face. "What? Who? What'd they get?" he finally got out.

"My laptop," Angela told him and started sobbing again.

Realization crashed in on Joe. "Did they..." he looked at the TV. The Xbox and Naqeeb's laptop were sitting placidly behind some packing boxes on the floor. The Nano agents were safe. Relief washed thru him.

"Did you recognize them? Who was it?' he asked Angela.

She sketched it out for him. "We walked into the apartment and saw this guy standing right there, she pointed. "Then the woman hit Naqeeb and the guy hit me and I, I..." She tried to hold back the sobs.

Joe took her by the shoulders and looked down at her. 'Look at me. What did they get? Think! What was on your laptop? There are no Nano agents on your laptop right?"

Angela focused. *Nano agents?* The questions clarified her mind. That's what they wanted she thought. "No! No. There are no Nano agents on my laptop. They did not get the CD we were experimenting on either. But my laptop does have all my notes and descriptions!" She lamented lastly.

Joe cackled. "So if they have a Nobel-level Molecular Materials Engineer and full facilities they might be able to reproduce your work in a year right? Fuck em!"

"Joseph!"

He came back to staring at her. "Sorry. Okay, sister, we are in some pretty deep shit here. Someone knows about System." And that someone wants them badly."

He looked at Naqeeb. "We got to move quickly, but smart."

"We have to call the police," Angela told him.

He pinned her with a look. "And tell them what? "We had a standard robbery? I don't even think they send someone out for that anymore." The home invasion and assault will generate some interest, but if there were no weapons? He trailed off leaving the question in the air.

She shook her head.

"Even if we file a report, the cops will never figure this out". He finished.

Joe moved into the living room. "The cops aren't going to go all CSI on this thing," he said turning back to her. No one is dusting for prints, or checking surveillance footage anywhere," He lamented harshly.

"But my notes", Angela began

Joe cut her off. "That's the thing. What can we do? If we say, Officer! You must treat this as a big deal because the laptop contained stolen notes about sentient Nano agents that are alive! See here?" Boom! We get arrested and the jack booted thugs swoop in and lock us away for 50 years." He stopped, breathing hard.

"Joe". Naqeeb said softly from the couch.

"Sweetheart, rest" Angela said returning and dropping down to his side again.

Joe came up next to the couch. Eyes closed Naqeeb fought to say something.

"Auntie Sarita", he got out. Condo."

Joe looked at Angela, puzzled.

"I think I know", Angela said. She stood, got her phone, and took some calming breaths.

She pushed some buttons. "Shireen? Angela. "What's up? Angela forced cheerfulness into her voice. 'I need a favor. Your Aunt Sarita is in India for the next three months right? Yeah." "Is she still trying to sell her condo?" Yeah, yeah that's kind of what I was thinking too!" I was hoping to surprise Naqeeb with the idea. "Just him and I in the city for two days!" No totally. No, we would never. We will. It's just that the real estate agent would never give me the lockbox combo. You do? Fantastic!" Joe started nodding along. "Tonight, if I can get him to go. Thank you! Got It.! Of course. I'll talk to you in a couple of days. Thanks again." Bye."

Naqeeb nodded weakly.

Joe approved. "Good girl." He frowned. "We got a place and we need to jump. How much cash do you have?" "We have to stay off the grid, no credit cards, no ATM and no cell phones as soon as we can." We can't use our own cars he mused, while Angela watched Naqeeb.

"Can you get a bag packed for you two? He asked her. I think we have an hour or so." *Less would be better* he thought but did not say.

He had no idea how long it would take for them to figure out the robbery didn't work. Angela went back to pack for herself and Naqeeb.

Joe dumped out a box from the living room and started stuffing the laptop, Xbox, and CD into it. Various cords and adaptors followed. His laptop and the microscope and Geiger counter followed. Various other computer CD's and items went in as well. He then raced back to get his clothes and toiletries. He changed his phone to make a call.

"Hey, Renee it's Joe. Honey, I need a big favor. I need a ride into the city in that huge SUV of yours. And I need it ASAP? Thanks, kid. See you in 10."

He checked on Naqeeb. His friend's eyes looked better focused. "Dude, I need you to be ready to move soon okay? Naqeeb nodded closing his eyes.

"Where is your Aunt's condo, Joe asked?

"Nob hill" came the weak reply.

"What? Nob Hill? Is there a doorman?"

"Yeah."

"Shit, we're screwed!" Joe said exasperatedly.

"No, I've got an idea if the room will stop spinning," Naqeeb explained. Joe just trusted and said okay.

"What's your cash situation?

"Grab it, Naqeeb said rolling to let Joe get his wallet.

Naqeeb had 64. His 71 plus Angela's 202 left them with 337 whole dollars. Joe also pocketed Naqeebs ATM card.

"We can make one last withdrawal, then were done," Joe told him.

They needed more cash. He went back to check on Angela. She was almost ready. She had to carry on bags ready by the door and she was putting things into boxes. Joe stopped her.

"Angela do you have any jewelry we can pawn?" he said slowly. She stopped and stared at him silently.

"Cheap stuff, please. They really nice stuff will just get us into trouble."

Angela started mumbling and pulling gold necklaces out of drawers and boxes. She got angrier with each one handed over to Joe.

"Keep that anger down please!" We need to be cold and calm and smart about this", Joe advised.

The knock on the door startled both of them.

"That should be Renee. Save the anger for the city."

Joe went out to open the door. He greeted the woman briefly, "sweetheart please just do what I say for a bit. We are in a little trouble here."

Renee looked at Naqeeb lying on the couch and nodded slowly.

10 min later they helped Naqeeb into the packed up SUV. Angela wondered if she would ever see the little apartment again.

"Head to that strip mall on el Camino Real Joe directed.

Convenient thing strip malls. This one adjoined a Walmart. First stop was the ATM. Joe withdrew the max for each card. 900 dollars later he hit the pawn shop with Angela's necklaces. 3,000 worth of jewelry became 345 dollars in cash.

Angela said nothing when Joe apologized.

Renee, Naqeeb, and Angela waited in silence as Joe took their hoarded cash into Walmart. He emerged 25 min later with a full shopping cart. The plastic bags went into the boxes and bags they already had.

"850 Powell," Angela told Renee.

Even at 2:00 the traffic was bad on the 101 heading into the city. Naqeeb lay on the back seat with his head in Angela's lap and an ice pack and towel over his face. Joe in the passenger side. Everyone was quiet.

He took off the towel and sat up when Renee turned up Powell from Geary. Hordes of tourist-swarmed the streets. Everyone thought he looked better.

"Turn right on California from Powell, he told Renee. As she turned the silver SUV onto California Street, Naqeeb ordered her into the buildings parking garage entrance on the right. There was a call box and keypad to open the gate. Naqeeb keyed in a number and the gate rolled open.

"We're in, Joe said. "That was easy".

"No, just the first step," Naqeeb told him. "Park in 901" he directed Renee. His Aunt's reserved spot.

As she pulled deeper into the garage and the assigned spot, the doorman emerged from the glass door to the lobby area. Naqeeb got out with a box. He managed a smile at the doorman. "Hey, Danny right?" He asked the man. "Good you got the message from my Aunt Sarita, okay?"

The doorman looked puzzled. He recognized the kid alright. He had seen him a few times with Ms. Sarita. They paid him to know who went where. And to make sure no one else got in. He told the kid that he had received no message.

"Really? Naqeeb said perplexed. "She called our house and said she was coming back from India sooner than planned. She told my mom to get her some food and bring some stuff out of storage." He gestured. "My mom called me and ordered me to do it." I was told you had been called." He set the box down and turned to Danny.

"You know my mom Aludra Al-Meri right? She's my Aunts emergency contact on the condo. You guys called us on that water leak last year?"

Danny remembered the kid coming to help his Aunt that time. He waffled.

"Yeah, I do but..."

Naqeeb immediately sympathized. "You should call my mom and verify with her, so you don't get in any trouble." Naqeeb held out his phone.

Danny smiled, "Nah that's okay. I know who you are." But there's a lockbox on the door, he said hating to be the bearer of bad news.

Naqeeb grinned at him. "That's cool the realtor gave us the combo." He motioned for the rest of his group to grab a box, which they did.

"I can't believe she's selling the place, Naqeeb told Danny.

"Yeah, I'm going to miss her, the doorman said. "She's my nicest tenant."

Naqeeb agreed. "She's heading to Rossmore, that big retirement place in Danville. He shrugged. Closer to family." Danny nodded. He understood that.

"We'll be out of your hair 20 min tops," Naqeeb said as he walked to the service elevator with the group. I can only impose on friends so much."

"Sure thing. Danny went back to the lobby and his video cameras.

The group crowded into the elevator. No one spoke as Naqeeb leaned against the sidewall breathing heavily. The doors opened on a drab landing. "The front elevator landing is much nicer," Naqeeb said gamely.

The lockbox yielded the key. They entered the kitchen. Naqeeb turned on the light and sat his box on the island. The white marble gleamed spotlessly. The whole kitchen did. They immediately made the second trip to retrieve the rest of the items in the SUV. They didn't want to chance Danny asking why they were bringing suitcases into the apartment. The last of the bags were put into fridge and island and the suitcases were wheeled into the main foyer.

Joe whistled. "Yeah, this will work."

Unit 901A occupied one-half of the top floor of the building. 2950 sq feet of old school Nob Hill San Francisco charm, met the four. Naqeeb went to lay down on the couch while the other three explored. They'd seen the gourmet kitchen. The entry foyer contained the same travertine tile and wood paneling. A small powder room and coat closet were on either side opposite the front door. To the left was the Kitchen entrance and the living room/dining room area.

His Aunt favored an Asian motif so screens separated the two spaces. The fireplace actually worked. The three bedrooms, two more bathrooms and a study that branched off the right side of the entrance area. Renee gasped as they entered the study. The room faced north directly along Powell Street. The window perfectly frames Alcatraz, the notorious prison. It gleamed in the sun.

"How much is she asking for this place?" Renee asked Angela.

"2.99 million and it's all yours, she said. Renee considered that amount of money.

"Plus the Condo fees to pay Danny," Joe put in. Do you think they designed this place around that window," he asked the women. They shrugged. The three wrapped up the tour looking in the bedrooms. The place was staged to sell so beds were there, but mostly the place was cleared out.

Angela went to sit with Naqeeb while Joe walked Renee to the service elevator.

"You going to tell me what's going on?" she asked.

"It's probably better that you don't know," he told her seriously. When are you heading to Santa Barbara?"

"Tomorrow."

"It would be better if you went down tonight," he said.

"How much trouble are you in?"

Joe said nothing to that. "If the F.B.I. shows up you tell them everything okay".

He gave her a huge kiss. "I'll call in a month or two if I can.

The service elevator doors shut on her concerned face.

Joe returned to the kitchen to help Angela unpack.

'She's okay?"

"Yeah".

They moved items into the fridge and boxes into the bedrooms and living room. They unpacked System but left the laptop unplugged.

Naqeeb again lay on the couch, albeit an upgraded version. Joe and Angela sat on the opposing chairs. No one spoke for a long time. The sounds of the city drifted up to them. A soft ringing double bell chime came from the street.

Joe smiled. "Is that the Powell street cable car bell?" He asked.

Naqeeb answered without opening his eyes. "Yes, it is."

"How do you feel, man?"

"Terrible. The dizziness is passing but it comes back if I move around."

Joe looked at Angela with concern. He mouthed "concussion?" She nodded.

"Dude, we may have to move you to the 15 days DL." That brought a smile/grimace.

He blindly reached a hand for Angela. She was out of the chair in a flash and kneeling next to him.

"I'm fine, just a little scared, she told him.

"Me too."

Joe cleared his throat. "Can we go thru it again?" We need some info to work with here."

Naqeeb's account was pretty basic. Came in, saw the guy and POW the lights went out. Angela gave the same summary as before: the woman hit Naqeeb, the guy hit her. They left and she got Naqeeb on the couch. Joe found them like that.

"Not more than 10 min had passed," she told Joe.

He led her back thru it asking more questions. "Where did you park?"

"The reserved space."

"Did you see anyone?"

"No, not that we noticed."

"Any strange cars in the parking lot?"

"Who notices that?"

Okay. Joe was quiet a moment. Angela go into the kitchen, please." She complied.

"Dude, what did the guy look like?"

"Black guy, my height. Ball cap. Sunglasses. I kept looking at the laptop in his hands and thinking What?!?" That's it, sorry," Naqeeb concluded.

"No, that's fine. Angela come back in please!" She sat again.

"Could you describe the guy?"

"He was black, a little shorter than Naqeeb, maybe an inch or so. About the same weight."

"Beer belly? Joe asked.

"No they both answered in unison. "He was in shape, Angela continued.

"I looked up from seeing Naqeeb get hit and he slapped me. Ohh, he was wearing surgical gloves!"

"Do you remember that Naqeeb? Joe asked.

"Sorry no."

Joe had Angela stand. He walked back about 10 feet from her. "So I'm the guy. You looked at Naqeeb and then?"

"I straightened up as he stepped towards me. He shifted the laptop into his left hand and slapped me with his right. He just kept right on walking out the door," She concluded.

"What was he wearing?"

"Giants ball cap and sunglasses. T-shirt and black sweatpants. And the gloves."

Joe nodded. He was in front of Angela. "You're what 5'3''? She nodded back. 110? Yeah.

"So 6 feet 180 lbs. in shape. Joe mused. He shifted mentally.

"Now the woman. Take me thru that part again." Naqeeb opens the door and goes thru first? Right?"

Angela agreed. "Yes, he was about 2 steps ahead of me going straight in. I turned a little left and thought about going into the kitchen first. We both saw the guy coming down the hallway at the same time." She went silent thinking. "I think we were both so focused on him we didn't really see the woman until it was too late." She kind of ghosted up next to Naqeeb and hit him in the head."

"Hit him with something or punched him? Joe asked.

"No, punched him definitely."

"Was she wearing gloves?"

"I, I don't remember she ground out."

"That's okay, that's okay Joe soothed. Did you see her face?"

"Briefly. It was mostly profiles. She never looked at me. Angela frowned.

"How tall was she?

"Tall! 5'7" or 5'8"!" Angela said

"Fat, skinny? Black or white? What was she wearing?" Joe pressed.

"White I think. Trim. In shape, she had on the same outfit a ball cap, sunglasses and sweats with a hoodie," Angela stated emphatically.

"Are you sure it was a woman and not a short guy? Joe asked again.

"No, her ponytail was sticking out of the back hole in her ball cap. Long haired guys don't do that." Angela said.

She went through the basic sequence again. "She came up and hit Naqeeb, stepped over him and then the guy hit me and they both left." Angela stopped. Upset.

Naqeeb slowly sat up on the couch. Angela went to sit next to him holding his hand and arm.

"What are you thinking Joe?" Naqeeb asked.

"Same as before: this is no ordinary burglary. Typically when they went thru a place in Hemet, it was electronics, guns, jewelry and tools. Things that can be pawned easily." Just like we did. He gestured to Angela's necklace. We got enough flyers from the cops to know all about it."

"But we surprised them," Naqeeb said.

"Yeah, Joe grunted. "Man, she knocked you out with one punch! One!" Granted a sucker punch but she didn't crack your head with a bat, or a tire iron, she hit you once. And she knew she didn't need another."

He paused for a second collecting his points. "And the guy? He never considered dealing with you. Like he knew the woman would handle it." No shouting, no instructions. He calmly let her sucker punch you while he pimp slapped Angela." Joe stopped, breathing a little hard, anger showing thru. He started pacing around the room.

"That says training. Military style training. I think both of them were wearing gloves. Do you really think it was a coincidence that Angela's computer was stolen?" If you guys hadn't walked in they would have got System and everything!" He stopped and slumped in the chair. The question is who?"

There were no ready answers to that.

Joe sat drained. The three friends tried to process what had happened to them. They sat mostly in silence. The day bled away while the three huddled in the living room. Beyond the food into the fridge, they didn't unpack anything. They didn't dare make much noise. They never turned on their phones or computers. Definitely no TV.

The stress of the robbery had left them all exhausted. Even Joe. The fight or flight high crashed in on them. They kept waiting for someone to knock down the door. Angela finally convinced Naqeeb to lay down on the bed around 7:00 pm. An hour later Joe told her to join Naqeeb as he would stand guard. She went gratefully. Joe fell asleep on the couch soon after she left.

Cantrell hung up the phone. Peg looked at him with a watery-eyed glaze. She was three drinks in at the end of a very long day. "That was Carmel. The laptop contains no Nano agents, sentient or otherwise."

Rage suffused her face. Cantrell rushed to mitigate. "It's not all bad. The laptop contains notes and files from the Chin woman detailing the Nano process. Dr. Cooper says the work is brilliant. Miles ahead of anything he has ever seen. Given some resources, he feels he could replicate the agents in two years."

Peg sat back in her chair. "We don't have two years. We have about 6 months she reminded him. "What the fuck happened to the Nanobots?"

She gave Cantrell a hard look. He blanched. "Eggert's team got surprised during the attempt." He held up two hands placating. "They grabbed the wrong computer it looks like."

Peg's rage was building. "I pay you and him so that surprises like that don't happen, she growled shaking the papers in her hand at him. "Did you see the stock is down another point today?" And the latest polling has me down 17!"

She slammed down the Chronicle. "And those bastards in Washington are reneging on their promise!"

Cantrell let her vent. The Republican Senatorial Committee had initially agreed to run some aggressive ads on her behalf. With her so far behind they were balking at throwing away scarce money on a lost cause. The Tea Party PAC's had spent money hitting Boxer hard as a Washington insider. The incumbent responded by running spots calling Peg an entitled extremist. That charge resonated with independents. The combination of the Scimitar launch disaster and the ads, hurt her poll numbers. Now the DC insiders balked. And the hole got deeper. She couldn't afford to be off message right now!

Cantrell tried again to calm her. "Eggert has people looking for them right now. We even have job offers in E-mails to them. The same dynamic holds: they can't go to the cops and they can't hide forever!" "We'll find them and grab the agents. Once we have them they can't do anything to us, he told her. "2 or 3 days tops!"

CHAPTER 12

June 5 3:30 pm 2010 Friday, San Francisco, Ca.

Hunger, boredom, and frustration drove him out into the streets. Naqeeb had to get out. They needed food and a cable connector to run from the outlet to the Xbox to the staged TV. He'd been holed up for the past 48 hours. The first 24 were mostly spent trying to keep his head on straight. That Wednesday afternoon and night was rough. Bluffing past the doorman had taken almost all he had. When he went down he went down hard. He slept till noon on Thursday. Waking with a dull headache he went to the living room to find Joe and Angela sitting, talking. They both jumped to their feet to see him vertical. Angela hugged him fiercely.

"How you feeling?" Joe asked.

"Better. I need a drink, an aspirin, and a shower in that order", Naqeeb informed them.

30 min later he sat next to Angela and looked between his best friend and the woman he loved.

"Who was it? The Feds?" he asked.

Joe shook his head. "I don't think so. The F.B.I kicks in the door and arrests everybody."

"Who then?"

"Kocinski" was all Joe said.

Realization pierced his head. Naqeeb turned to Angela. "I'm sorry honey. You didn't want to go to him in the first place and he sold us out."

Angela held his hand. She shook her head. "No, we all agreed. This is his fault not anyone of ours!" She said with finality.

A quick peck on the cheek from Naqeeb. "That's my girl. Then added the big question: What do we do now?"

Joe stood and faced his friends. "I see two options: One goes to the Feds and Mea Culpa and face the consequences and let them handle it. Or…"

"Punch the bully in the nose!" Angela said surprising them.

"Exactly!" Joe grinned at her.

Naqeeb urged a little caution. "Whoever is ultimately behind this we know it wasn't Kocinski acting alone, right? The others agreed. "So what about our families, Naqeeb asked.

The three debated the point. Argued really. The general consensus solidified around the notion Joe had: "They are probably watching our families looking for us. That's why we can't contact them. But they also know any act to hurt them drives us right to the FBI." "They want the Nanos. Joe said passionately. And they don't think there is a damn thing we can do to prevent them from taking them!"

He studied Naqeeb and Angela in turn. "We have to figure out who it is, and hit both Kocinski and them to get them to back off."

He got agreement from his friends. They got down to work. Unpacking the Xbox and the Laptop they dragged the coffee table closer to the TV. They put a sheet down to protect the table. The three looked longingly at their cell phones while they setup System.

"Don't even think it Joe warned. "We might be able to risk the drop phones he told them. Let's wait a bit." Part of his Walmart spree was pre-paid cell phones.

Booting the laptop, System immediately noticed the configuration change.

"Where is the display device"? "The internet portal is different."

Angela typed in response: "The laptop will be the primary display device until the connector is obtained for the TV."

System acknowledged the change easily. "More efficient, System acknowledges."

Naqeeb hummed a little.

"What?" Joe asked.

"More efficient for them to display but harder for us to process." Angela started in explaining how they viewed the two screens.

Joe and Naqeeb also spent some time discussing how to present their current situation to System. The danger was not something the Nano's could comprehend. One system interfering with another systems operations? A prioritization of needs and process hierarchy? That System could understand. Interfering so much that creator Joe Angela or Naqeeb were sacrificed? That could not be allowed to happen. The creators must be preserved. Angela felt a wash of affection for the Nano agents.

Joe got down to some hard questioning. Could System obtain information on system Kocinski with no other system finding out? No other system could be allowed to detect the presence of System? It took hours to define parameters, what types of information and how to display the data.

Joe finally straightened from his data sheets and the laptop. It felt good to be doing something to rectify their situation. Anything felt better than sitting in a huddle!

He was angry. Angry at himself, and Kocinski, and at whoever had hurt Naqeeb and Angela. Revenge was a dish best served cold. They said. Joe was running a little hot right now. To ensure they could get a little payback they needed to ensure no one could track them back here. It came down to a want of a nail kind of game. Joe wanted revenge and to inflict a little pain on whoever did this. To do that they needed info. To get that info they needed to hook System into the internet. Securely. How do they do that?

The three finally decided late Thursday afternoon to risk quick calls home on the drop phones from Walmart. "Busy packing and getting ready to move! Maybe between places for the next few days, call you soon!" The families had no suspicions.

By Thursday night Joe managed to get his points across to System. The final message displayed on the laptop was for the creators to obtain a hard line cable connector versus the wireless remote in the condo. System could mask its activities better with the hard line. The three planned into the night.

Naqeeb had to wait until Friday afternoon to get out of the condo. Danny went off shift at 2:00 pm. Danny finally left and the weekend people took over. The building came a little more alive in the Friday sun. People naturally went out more, did more errands, and ate out more at the end of the week versus a random Tuesday.

Finally, at 3:30 Naqeeb now walked along California Street down the hill to China town. Turning left at Grant he went thru the gates like any other tourist. The Canon Source store provided the cable and connector he needed. He picked up a few other odds and ends. Three doors further down was the Far East café. 75.00 bought three big bags of Chinese food to lug back up the hill. At 4:19 he opened the car gate back into the building. As he entered the garage, the service elevator opened to reveal Joe waiting for him. Joe grabbed two bags and they rode up in silence. Angela was waiting for them at the kitchen door when they arrived at the ninth floor. 90 seconds from street to condo Naqeeb thought. Smooth.

"You got a beer?" Joe asked, putting away food.

"The guy asked. He insisted really and it went with the cover of ordering enough food for a party," Naqeeb reasoned.

Joe grunted. Naqeeb handed off the cable connection to Angela while he helped Joe.

When they emerged from the kitchen onto the living room Angela had the connection established and System was off and running. "Searching within defined parameters," System displayed.

Angela typed. "System will display results one hour before cycle off."

Naqeeb crooked an eyebrow at her. "If they have something definite we can adjust and have them focus on whatever they find," she explained. If they find nothing then we can spend the night redefining the search parameters."

"Good thinking."

They settled in to wait. "Anything out of the ordinary, outside? Joe asked Naqeeb.

"Nothing I could see, he answered. The worst thing was coming up California Street. That hill is a bitch."

"How's the head, sweetheart?"

"Better," Naqeeb responded to Angela. Everyone was relieved that Naqeeb seemed to have no lasting effects from the blow to his head.

They lapsed into silence and waited. They ate some food and waited. They hooked up Aunt Sarita's printer and waited.

"Dude, Joe said, good call getting some paper and a print cartridge."

At 7:00 pm System Ready was displayed. Joe looked at the other two. "Here we go."

He typed "display results." System started displaying 700 pages of data on the screen. "Shit," Joe said. Go get a CD we can dump this too, Joe directed.

Naqeeb winced, "Sorry, didn't think of it." He went to the study to scrounge one from his Aunt.

Joe typed to direct System. "System, recompile results and place into four areas of concentration: Personal, Financial, Work, and all others." The screen went blank for a second and then: System acknowledges."

Naqeeb returned with the CD. A blank screen for a minute while System sorted the items then, System ready was displayed again.

Angela suggested that System displays the personal items while putting all the info on the CD. "You and I will go over the screen display while Naqeeb prints out CD info."

Yes from all involved and Naqeeb started printing out pages of data.

Joe and Angela put their heads together over the personal data. The data was haphazard and in mind-numbing detail. Kocinski's marriage seemed to be the largest file. The divorce decree and the transcripts filled over 100 pages. They took 20 min or so and skimmed thru the file.

Joe rubbed his eyes. "This is getting us nowhere."

Angela agreed. 'Where should we look?"

"Follow the money. Isn't that the old adage?" Joe told her. She agreed and they started thru the financial data. They found the first clue 10 min in.

Naqeeb looked up from Joe's laptop and the CD. "I'm starting to print out the financial stuff"

"Good, Joe agreed still reading the screen. His brow furrowed. He pointed to something to Angela. "Find this name again." She scrolled thru the data until the name popped up. "Here". Joe read thru it. "Okay, I got that". He came to the printer and sorted thru the offerings. He picked out 10 or so pages and went to the dining room table.

Spreading them out he checked his watch. 14 min to 8:00. Shit. He looked at Angela. "I've got something but I'm not sure what it means." Naqeeb and Angela joined him at the table. "Fastfac" Joe told them pointing. "It ends with them."

He checked his watch again. "Let's do this: Let's reset System to look for a FastFac link. But we only give them an hour to look over FastFac. That way we have time to look over this material, get System working on what we think is the best lead but still can redirect if it's a busted path."

He looked searchingly at his friends trying to see if they agreed. Angela nodded and Naqeeb joined her. "That's a good idea, Joe, he said. "I'll type in the new instructions for System." Angela had a suggestion: "Why don't you just tell system to search along the same lines as Kocinski, substituting FastFac for the individual?"

Naqeeb agreed immediately. "And I'll tell them to categorize the results into the same areas." Both Joe and Angela thought this was a wonderful idea.

Naqeeb typed away for 5 min or so. He beat the cycle down a deadline and put System to bed. He wondered for the umpteenth time if this was what parents did: get the child put down and then steal a few hours for themselves while the kid slept.

Naqeeb got up from the computer and stretched. Going into the kitchen he came back with a takeout box and some redbulls. The three passed around the food and munched in silence.

"Okay, let's see what you got," Naqeeb told Joe when they finished.

Joe bent over the table pointing at the papers. "This is the smoking gun, I think. Professor Larry Kocinski in his capacity as CEO of Lifestyle LLC signed a personal services consulting contract with California media partners for 10 million dollars."

He had their attention. "California media partners is a shell corporation that seems to do all its business with PullPac. "Pull Pac is Margaret Pullman's campaign financial engine. PullPac buys all the media and campaign items." "Pullman owns FastFac."

"Look at these, however: Lifestyle has used that money to buy a huge amount of FastFac stock."

Joe took them thru brokerage statements that detailed the very recent stock purchases. Kocinski had eight brokerage accounts with various stock brokers. UBS, Goldman Sachs, JP Morgan among others. "He was pretty smart about it. He had the money in 750,000 to 900,000 dollar chunks all invested in FastFac stock."

They went thru it a few times: Kocinski to LifeStyles to Cal Media to Pull Pac to FastFac to Pullman. Angela had trouble seeing it at first. "He got 10 million from them ultimately, why dump it back into FastFac stock? That stock is worthless, right? The Scimitar isn't very good you said." Won't he lose money?"

Naqeeb answered. "What's that stock going to be worth if FastFac announces they have sentient Nanotechnology?"

That simple statement stunned them for a moment. Joe whistled. "He got greedy. The stock is going to explode. Those are the kinds of trades the SEC scrutinizes."

Naqeeb disagreed. "I don't think so. It's the same as any other employee investing in the company. What about all those FastFac workers who have a company stock plan?"

"Still, he got blinded by the dollars", Joe said.

"We need to figure out the connection between Kocinski and the company. Naqeeb said. "Maybe it's a personal relationship with Pullman", he theorized.

Angela was quiet for a few minutes thinking. "Was it FastFac that robbed us she asked, getting upset.

Joe considered and nodded. "I think that's likely. They might be blinded by the dollars too. Stanford wasn't amazed by your work but someone sure thinks it's valuable!"

They started typing parameters for later input into System, shifting over to FastFac as the target. Naqeeb wanted to get System working on the link and any info on FastFac while they went through the rest of the Kocinski material. Joe especially wanted to see if they could tie Pullman and FastFac to their robbery. Angela was a little skeptical on that. "There is hardly an interoffice e-mail that says: please go rob apartment 210E at the Housing Union."

"You'd be surprised what gets written down. Joe told her. "Plus it's all how corporations pay for those kinds of things. Look at the personal services contract with Kocinski. Follow the money, follow the money, follow the cash!"

They worked in silence. Things were starting to get a little murky. "What are we going to do once we have established that FastFac is responsible?" Angela asked.

"Then we punch that bully in the nose! Joe said with conviction. Angela wavered.

The three friends debated options.

Naqeeb brought up a telling point. "We cannot stay here forever. We need to back them off so we can get back to our lives! I figure we can only stay here another day tops." He stretched and got up from the table bending backward to flex his back. Sitting hunched over a computer screen all day killed his back. He continued making points. "We are going to need more food and money. It's no secret this is my Aunts place." "If they are looking for us we need to be gone, Monday am at the latest, Joe." Naqeeb finished up.

"You get no argument from me, Joe told him. "Let's move like we got a purpose," he told them.

The friends dug into Kocinski's files. His scandal at work was laid bare. His divorce, and finances all showed how he was living.

"Did you know about the affair with the student?" Angela asked Joe. He shrugged. "It's a small campus. There's always stories going around. I never really believed them," he admitted.

Naqeeb agreed. "I knew he had had some trouble but the University did a good job of hushing it up. "And look at this list of published papers. The guy is smart", he concluded.

"He figured me out pretty quick, Joe griped. A quick check on you two and he made 10 million toot sweet. And he stands to make millions more." He was fuming.

"He needs it Naqeeb said. "Look at his bank statements. The guy is pretty close to broke. A gleem came into Naqeeb's eye. "He's not "Joe Broke" but he is hurting, Naqeeb teased.

"FU," Joe shot back, smiling. Nice to have a little of the old Naqeeb back.

Joe glanced at the bank statements for the corporation Kocinski had setup. "Christ look at this. He pointed to the address. "Commercial Union building". Half of the brokerage accounts listed an address in that building. He got up from the table. "Follow me." He marches the couple into the study. He pointed out the window. "See that square kind of black wavy looking building right there. That's the Commercial Union building. Half of Kocinski's money is in that building. Shit, Wells Fargo is two blocks north, and JP Morgan is in the Trans-America pyramid". Joe pointed out the iconic SF skyscraper. *Christ the money that flows thru this town* he thought.

He looked at his two friends. "The money will damn FastFac too." "Fucking Kocinski, he said.

Naqeeb quietly said. "That's right fuck, Kocinski. We take that money from him."

Joe whopped!

"Can we even do that?" Angela asked Naqeeb.

"Hell, yes. Maybe. I don't know, Naqeeb admitted. But he forfeited that money the second that guy hit you!" He hugged Angela.

Joe put a hand on his shoulder.

Naqeeb looked at his friend "We got plans to make." And make them they did.

CHAPTER 13

June 7 5:30 pm 2010 Sunday San Francisco, Ca.

The cab pulled up on California street side of the building. It double parked. The three young people emerged from the parking area carrying boxes and hauling luggage. They loaded up and piled in the cab and took off. The lady sipping coffee at Café Centro on the corner of California and Powell had the perfect spot.

She could clearly see the cable cars and up and down Powell as well as the front and side of the Aunts building. The city was basking in perfect June weather. 71 and cloudless. The sun would be out until 7:30 that night. Sunday seemed to be tourist swap out day. One set leaving and one set coming.

The three almost blended well. But she was looking specifically for them. The woman, code name Winter, took note of the cab number and the license plate. She was a thorough professional. She was calling it in before the cab turned left on Stockton.

Angela was freaking out in the back seat. She kept chanting to herself: "This had better work, this had better work." The cab pulled up in front of the Hotel. The neoclassical façade was intimidating, to say the least. Joe handed the driver a twenty and told him to wait. Angela stayed in the cab.

Joe and Naqeeb allowed the bell hop to take one suitcase each and let them lead the pair in. The columned front swallowed them up. The setting sun making the building glow. They approached the front desk with confidence. The young man was most helpful. "How may I assist you today, gentlemen."

"We have a reservation for two rooms. Should be under Singhe, Jatinder Singhe. Jack really," Naqeeb said smiling.

"Certainly sir. Is this your first time with us at The Ritz Carlton San Francisco?"

"It is."

The receptionist typed away. "Yes, here we are. He studied the screen and then smiled broadly. Yes, Mr. Singhe, we have you and Mr. Roberts..." he looked at Joe expectantly who nodded and smiled. "With us for a week, he finished. "Two adjoining Garden Suite rooms, very nice. And, I see the rooms are paid for."

He stopped and looked at Naqeeb. "May I see it?" He asked conspiratorially.

Naqeeb handed over the UBS Thumb drive nonchalantly. The receptionist reverted to a 22-year-old tech geek. "Cool. We just started accepting Bitcoin on 1 June and this is the first time I've even seen one."

"Sure, sure, said Naqeeb graciously. Plug it in."

The receptionist did just that with the digital wallet. Naqeeb's picture showed up as an ID point along with his vital info. The most vital fact being he had 67,847 "dollars" on this wallet.

"How does it work?"

Mr. Roberts took over. "This is the wave of the future he said knowingly. Nobody is going to have money or unsecured credit cards anymore." Think of it as a combo of a foreign currency and a debit card."

The young man thought through that. "Each transaction has to go to an exchange where the current bitcoin value is assigned. Then it's just like a debit card. The value is subtracted and you get whatever you pay for." "The hard part is that if the value of the Bitcoin goes down, your purchasing power does too, just like when you are in Europe," Joe said being an asshole.

Naqeeb cut in smoothly. "No, the hard part is writing the check to put the money up in the first place."

The receptionist laughed over that. "May I set up an account for incidentals on the rooms, Sir?"

Naqeeb thought that over. "That would be smart he said. Put 5,000 in each room in case we need anything."

"Certainly."

He motioned to the bellhop. "I will have your bags delivered to the rooms' sir." Is there any more luggage?" inquiring.

"Not right now said, Joe. We are going to head out and see a little of the city before we have a late dinner."

"Excellent sir. If you need reservations or tickets to anything while you are here please see the concierge, she will be happy to assist you."

"Thanks, Kevin, Naqeeb told him.

"Enjoy your stay!"

They exited the front and slid into the cab. Naqeeb winked at Angela, who let out a huge breath. It worked!

"No rooms? The cabbie wanted to know.

"The intercontinental, Joe told him without answering. The driver went south of Market and pulled up in front of the giant blue glass hotel a scant 5 min later. Another twenty to the driver. They were gone exactly 17 min. The cabbie stopped asking questions after the Sir Francis Drake. They even got rooms at the Fairmont. The Fairmont hotel where Café Centro was located. Ms. Winter had departed 47 min earlier.

She would regret missing them.

Now the cabbie got one final twenty to drop them at Union Square. Boxes and bags in the passenger's hands as the cab drove off. Shoppers, tourists, homeless people, cable cars busses, and pigeons all swirled around Union Square.

The final Hotel loomed across Powell. The Westin St. Francis. It was easy to get lost in the crowd when there are 1100 rooms in the building.

Checking in with Bitcoin had lost its luster. The St. Francis had lost none of its. The adjoining suites faced east on the 18th floor. The bay bridge was in all its glory. They met in Naqeeb and Angelas' room. They needed to get things and plan.

"This is Mr. Singhe, in room 1827. I need a reservation for the steak place in the lobby. Yes, Bourbon Steak. 8:30? Perfect." Angela looked at him. "When in Rome babe!" She rolled her eyes.

Joe tried to save his friend. "Hey, we can't help it if only 5-star places are accepting Bitcoin!"

The virtual money and digital wallets were simply no match for System. Once the concept was explained to them The Nano's took 27 min to "allocate" Mr. Singhe some money. Some 67,000 odd dollars. A quick picture from one of the smartphones and Jatinder Singhe had plenty of money. In a digital wallet.

Bitcoin worked on 32-bit encrypted pairs assigned as tag ends of the transactions. The Nano agents were simply fabricating both pairs to start off Naqeeb's account. The tricky part would be when Bitcoin started an internal audit. System told Joe they were only auditing on a quarterly basis. Sloppy.

The three needed to fly from the condo and one of the downtown hotels was the best option. The digital wallet and bitcoin fell into their laps at Systems suggestion.

"The creators should allocate some assets from the exchange. "Is that not what it's for?" System wanted to know.

Uhhh. Kinda. They prepped and went for it.

Now 5 hotels later sitting in the Westin, Joe felt Kocinski's money didn't stand a chance. They had to move in quickly but could they get all the paperwork in place in time. Tricky. They had to place sell orders with all the brokers. The sell orders would be executed at Market open on Monday. Market close, the money would technically be available. That was 1 pm PST. How to get at it? The signature authority on those accounts was Kocinski only.

Could they get Ms.Satoshi Nakatomi added? System assured them they could do it. Can System monitor the communications between the broker and Kocinski to make sure nothing happened? Yes.

Could System place the requisite forms in the computer files to ensure the money would be transferred. Yes. But System wanted to know why? Would it not be simpler to create new money in the accounts for the creators to use? The idea is to take the resources from Kocinski and give it to themselves, they explained.

"That's a polite way of saying stealing" Angela muttered while typing.

They faxed backup forms to banks and brokers on Saturday. The forms needed to be processed by 1:00 pm Monday. They should be. *Should be.*

There was no way System could stop the regular mail notifications from going out. Wednesday or Thursday, Kocinski was going to know something had happened.

"Fuck him Naqeeb said. He doesn't get to profit."

The three rode the elevators down at 7:00 pm, leaving System hard at work. Freshly showered. Joe loved the soap the hotel was using. His skin was so soft!

The Westin is a full-service hotel with many fine shopping opportunities. They needed clothes to look the part for the business tomorrow. Angela went conservative. A little black cocktail dress and shoes. She also bought the woman's version of a business suit. Donna Karon had a lovely skirt and Jacket combo. Navy Blue with a cream colored blouse. She needed accessories. Two thousand one hundred sixty-two dollars later she emerged with her hair and makeup done looking fab. Would Ms. Nakatomi like the rest of her things sent up to her room? That would be lovely.

She met the transformed men in the lobby. She gave Naqeeb a huge hug and kiss and wiped lipstick off his face. His hair was trimmed and styled and he wore a gray jacket with a tailored white button down shirt open collared. The matching gray pants were cuffed and the shoes Italian.

Joe was similarly shod and dressed in black. Angela complimented him.

"These socks cost 38.00 dollars he told her. All three giggled going up the stairs to the restaurant. The place was dark and only half filled. The corner table allowed them to review.

"The first one will be the trickiest, Joe told Angela. You have printouts from the Lifestyle's statements System copied." Let the account manager see that you have those. You should ask to get a copy of the sell order and the confirmation, then ask for the withdrawal. A cashiers check," he said softly.

Angela sucked in a breath. "That's going to raise eyebrows, she said.

"Agreed, chimed in Naqeeb.

Joe nodded. "If they give you any static, ask to see the manager. Remember you are not closing the account. Leave some weird number over 50,000 in the account like there is a very specific number that you need." Joe paused as the waiter brought salads. He leaned into Angela. "If they ask why, tell them it's none of their business. You need to act like it's your money."

Naqeeb chewed and said around a mouthful of arugula, "Time is going to be our real problem. We have to be done by 5:00. There are 6 accounts down there we want to hit: Four in the CU building, one at the Wells Fargo main branch and JP Morgan in the Trans America building."

Angela and Joe nodded along. "we only have about 35 min at each stop, Joe said, that gives us 9 min to go from building to building."

Angela still looked doubtful. Naqeeb sipped his wine. "Honey, I'll be right with you the whole time. Plus I'll be on the phone continuously with Joe while System monitors the transactions to alert us if anything goes wrong."

Joe smiled. "Piece of cake." "Once they start drawing the check, send Naqeeb to the next appointment. After the last one at the CU send him down for the cab. The cab should be out front at 2:53, waiting." 2:53? "That's what my program and Systems says, Joe replied to her.

Silence ruled as the waiter brought the entrees. Angela eyed Joe's dish. "What did you get?" The waiter was setting down a copper dish and began preparing the meal tableside.

"Lobster Pot Pie," Joe told her.

"We are off the deep end now", she thought.

They decided to share a dessert as the sommelier opened the second bottle of wine.

The Grand Marnier soufflé was excellent.

"Okay, Joe said a little drunk. Contingencies? Problems?"

"We need some real money, Naqeeb said. This is nice, he waived, but we only have around 600 left."

"Can we cash in some bitcoin?" Angela asked.

"I think so, her boyfriend said. There is like a daily limit, maybe a thousand or so? I don't know."

"You check the front desk when we go up, Joe told him. Then we spilled whatever we have in case we have to bounce."

Angela had one last freak out. "What if the brokers call the cops?"

"They won't. You have ID, Joe reminded her. The state issued ID card was fake. It used the new name and Angela's data. "Plus, he said sounding confident, the broker has your picture and signature on file. That's what it exists for in the first place!"

"What if they ask to see a second form of ID? A driver's License?" She persisted.

Naqeeb took over "Honey, the copy of the faked passport is in their files per System. You don't have a driver's license because who can afford to park in the city?" Joe quizzed her: Where do you live? 850 Powell unit 901A. Exactly. Social? She rattled off hers with the last four reversed. Date of birth. Again day and month reversed. Angela recited correctly. It was a risk giving information that could trace back to them, but they had to take some risks. The men felt she should stick as close as possible to the truth.

"Time is the enemy, Joe said again. If they stall and say they can't cut the check until someone else signs it or something, I'm not sure what to do. You'll have to play that by ear.

The friends exited the restaurant and Joe and Angela lingered by the elevators while Naqeeb hit up the front desk. He returned with a small satisfied smile. In the elevator car, he passed out a few hundred apiece.

Joe cautioned them again. "Stick to the plan, and the timetable. Where do we meet?" Intercontinental at 6:00 pm."

Joe nodded. Three hundred each won't last long in this town. If you have to run, goodwill is on Market. Try to change hair color. The youth hostel is on Mason. We have reservations under the new names."

They paused by the room doors. Joe looked at his friends. "We meet at 8:00 am to get System up and working."

"Okay."

"K."

"Get some sleep."

CHAPTER 14

June 8 0730 2010 Monday Mountain View, Ca

The TV crew was setting up in the conference room. The local station's political reporter was getting an exclusive in-depth interview with Margaret Pullman today. He hoped the network would be pulling some of this feed for their broadcast. The cable affiliates were salivating over the footage. Pullman was a polarizing figure today after her primary win. Was she an elitist billionaire business woman trying to buy her way into office or was she a no-nonsense outsider who could clean up Washington? He had hours' worth of questions.

Cantrell and the press liaison hovered in the back of the room. This needed to go well. Today was the first day of the pivot. The general election started today. This interview would dovetail into a major ad buy to introduce Margaret Pullman to the general electorate. The ads would show Peg as the nose to the grindstone leader in the cutthroat world of high tech.

FastFac was a 3 million dollar a year business when she came on board 25 years ago. Today it was a fortune 100 worldwide leader in innovation and customer service. She was exactly what was needed in Washington to shake up the establishment and bring a voice for the people. Peg Pullman. Leader. At least that was the plan.

Cantrell left the conference room and went to Peg's office. He found the star of the ads drinking vodka. She was breathing hard, angry looking at Eggert.

"Goddamn, it finds them!" Cantrell pulled the drink out of her hand and handed her a mint.

Eggert handled the rage smoothly. "We will, he assured her. We tracked down the cab last night. We have it narrowed down to 4 hotels in the downtown area. It takes time to get the numbers of people needed to cover the large hotels." He listed the places.

"How can they afford the Ritz? Peg asked.

"The girl's family is pretty well to do. And connected. Her dad has a meeting with Ed Lee today as a matter of fact. We think they are using some of that money." Eggert went over his notes.

Peg knew who the city councilman was. Bureaucrat! Who cares?"

Eggert nodded. "People run along predictable lines. Patterns. Especially under stress." Eggert looked at Cantrell significantly. "These three have been smart. They took computers and phones with them from the apartment and haven't been back. They've stayed off the grid. Smart. However, they had to run somewhere and they went to family. Predictable." He looked at the two. They bolted and tried to cover their tracks. Smart."

"Can't we force the families to tell us where they are?" Pullman spat.

Cantrell broke in. "That's not a smart call Peg. We can't really touch their families. If we do they run straight for the feds and its game over."

Eggert picked back up. "We have them run to ground in one of those hotels", he reminded them. "I need to get more people."

Peg snarled around the candy. "Fucking idiots! Three kids. Find them faster!"

Eggert seemed reluctant. I'd have to bring Porthos and Artimus back from Carmel."

Cantrell cut off the blast from Peg. "Just do it. So what if they've seen them? They can't tie those two to us."

Eggert agreed and said they'd be on scene in four hours. He left the room.

Peg composed herself and left for the interview. She secretly hoped those kids saw her people. Scared people make mistakes. They don't think rationally.

Angela sat in the chair and sweated. She kept glancing at the clock behind Mr. Zauhar. The account manager altered between a slight frown and a smile. *He knows something is wrong, but he can't put his finger on it,* she thought. Naqeeb's hand on her back soothed her. Whatever the investment firm had done to try to contact Kocinski, Joe and System had apparently figured it out. Hopefully. She was glad Naqeeb was cool handling this. Naqeeb kept nodding and murmuring "okay" on the phone.

Naqeeb listened to Joe's voice on the phone. "14 min to go. I've sent the response through System. Kocinski has an appointment with Zauhar on Tuesday at 4:00 pm." Naqeeb murmured another "Okay," and touched Angela.

"Ms. Nakatomi, I've been blind CC'd on the e-mail Mr. Kocinski sent Mr. Zauhar. They have a meeting on Tuesday at Four. That will not meet today's payroll problem." He reminded her pointedly.

Angela straightened. This guy wasn't as tough as Aludra was he? Not half as tough as my mom! She pinned him with a look. "Mr. Zauhar, Mr. Kocinski is the CEO of Lifestyles. I am the CFO. I have cash flow problems he does not seem to want to deal with. I told him we would have to use the FastFac money to meet those obligations. My dealings with him are my problems. You have my documentation. Now produce the check or a statement why you are declining a signature authorized person on the account. I have another appointment in 12 minutes."

Silence. Mr. Zauhar settled on the smile. "Of course Ms. Nakatomi. We must do our due diligence on these matters", he said. "We can have the check drawn up immediately. 697,000?"

'Yes."

"It will take another 20 minutes or so I'm afraid, Zauhar apologized.

"I'll send Mr. Singhe back to retrieve the check from your receptionist, will that be satisfactory? Angela said. Again the slight frown.

"Would you like me to do it myself?"

Zauhar exhaled. "Yes that would be better, I think," he said.

"In 20 minutes, then", Angela said and shook hands and led Naqeeb out of the office.

They stumbled into the elevator thankfully alone. Joe informed Naqeeb that UBS was frantically trying to get Kocinski on the phone. System was routing all calls from them to Stanford or Kocinski's home to telephone purgatory. He wouldn't be getting any calls today. Or texts. Or e-mails. He was electronically isolated. Kocinski was having a bad day and wouldn't even realize it until Wednesday.

Angela sheltered in Naqeeb's arms as they rode up to Charles Schwab. That was the third stop on the itinerary. Goldman had been a piece of cake. 722,000 dollars' worth of cashier's checks nestled in her portfolio. Yes Ms. Nakatomi, of course, Ms. Nakatomi. 35 min from start to finish.

Now Schwab. She was getting better. They forked over 602,000 in 31 min. They rode back down to UBS and grabbed the check from them, before heading all the way over to the Morgan office at the TransAmerica building. They had 29 min to go.

Joe fretted as Naqeeb whispered the various questions and roadblocks the brokers were putting up. System was growing much more sophisticated as the Nano's got used to the way Joe wanted information presented and what the other systems were doing to verify their paperwork. Joe was seeing graphical representations of the brokerage computer trying to contact Kocinski.

He also "saw" which files the computers were accessing to verify their documents. Joe was amazed. It came down to computer protocols and certifications.

System kept providing information and requests as promptly as it could. Whenever the brokerage computer went to another computer system, such as the states driver's license database, System would intercede and mimic that computer and acknowledged the request and provide whatever data the computer wanted. Of course, that info was faked!

Joe learned an awful lot about computer systems that day. Computers don't "read" documents. All they did was verify them. Thru checksums and data fields and registers. Drivers licenses, birth certificates, forms were just ones and zeros to computers, as long as they added up right, the computer didn't know any better.

Take Ms. Nakatomi's birth certificate. That form was just data fields, ID markers, and formats, with checksum digits. They had to plant a birth certificate to establish Ms. Nakatomi as an authorized signatory on the account. The brokerage computer just tried to validate the certificate with the states ID database.

System mimicked the database and provided the ID marker back. The form was accepted into the brokerage computer as valid. System then went in and updated the form with the faked data. Voilà a valid birth certificate checked with the state.

No human being looked at the "date last modified" stamp. The computer didn't mind another system looking at the file and approving it. Someone was supposed to validate the file right? The broker's computer expected that users would access the data. System did just that.

The display Joe was watching showed System monitoring all the data flowing in and out of the CU building and responding whenever Schwab or wherever Angela and Naqeeb were raiding, asked to talk to another entity regarding the account. On a hunch, he typed into System. "How much of System is being used to observe the data flow from the target?" System responded that 42,738,620,537 units of System were engaged in working with Creator Joe. He relayed that information to Naqeeb. The "Okay" response was hushed. Schwab was fine with Angela taking the money, just slow getting the check ready. They waited it out with Naqeeb downstairs holding the cab.

The cab ride to Morgan was silent. Joe was tense in the Westin watching. Morgan went smoothly. Angela was more practiced as she heard familiar questions. The good people at Morgan handed over 738,000 dollars to Angela. Taking the check to join its friends gave Angela a small bad thrill.

One more to go. They got to Wells Fargo at 3:47. They went in hopeful. Ms. Williams the account manager listened politely and went out of the office to verify all of the documentation. Joe thought they were good to go. Ms. Williams reentered and stared at Angela.

"Ms. Nakatomi I have reviewed all of the forms and you are a signatory on the account. I got a call from Mr. Kocinski about 45 min ago. He was having problems with his credit card. Our banking side alerted us. He said nothing about any of this."

Ms. Nakatomi froze a little. Naqeeb whispered questions to Joe. "Shit! He said. "I see it now. He must have called from a landline somewhere." We didn't setup for System to monitor his credit cards and allow certain purchases, Joe concluded. My fuck up sorry."

Ms. Nakatomi unfroze. "You've spoken to Mr. Kocinski today? Funny. He seems to be avoiding my calls. Try again" she commanded Mr. Singhe.

They waited while Naqeeb dialed. He announced that the line was just ringing and ringing.

Ms. Williams said, "Yes, Mr. Kocinski mentioned that he was having bad phone problems."

Angela tried the hardball payroll line again.

Ms. Williams leaned forward in her chair. "Ms. Nakatomi you are free to close out the account here as soon as I speak with Mr. Kocinski about this. As I said you have all the proper documentation, but this situation is *unusual*. I've explained my thoughts to my vice President and he apologizes as well. But we would feel more comfortable issuing the check to you after we have verified it with the person who opened the account in the first place."

Angela tried one last gambit. "And if I sue".

Ms. Williams shrugged. "If this all checks out I'll be fired."

Angela rose. "I will try to get Mr. Kocinski down here to arrange the transfer or at least call you to set it up. Would that be acceptable?" she asked. Of course. They left without looking back.

Naqeeb informed Joe that Wells Fargo was a bust.

"Stick to the plan," Joe told him.

The main lobby was crowded with people using the bank. Naqeeb and Angela stepped to the center island. Angela opened the portfolio and dumped the bogus account statements into the trash. She placed the 5 cashiers' checks and a long two-page letter to Kelly into an envelope. Naqeeb sealed it. Stamp and address already affixed. They slipped it into the mailbox that Wells so thoughtfully supplied to its customers.

Stepping into the late afternoon sun on California Street the pair hailed a cab. Neither spoke beyond telling the driver to take them to the Intercontinental on Howard.

The left on Stockton brought them down the east side of Union Square. Naqeeb hoped Joe was following the plan.

He moved as soon as Naqeeb told him Wells was a no go. Stick to the plan. He put the three newly purchased roll on bags on the bed. Their clothes and toiletries were inside. The one bag he carried had all of their real valuables. System and the Xbox. Their real wallets and phones in addition to a growing arsenal of electronics. Two more digital wallets, drop phones, cords, and thumb drives all crammed into the bag. Each of them now had bitcoin wallets in their real names and fake. Two new drop phones had joined the party. Christ! He was forced to buy a label maker and label the electronics to keep them all separated!

Joe left the room and rode the elevator down to the lobby. He paused at the front desk to pull out 300 real cash from the digital world. "Enjoy your evening Mr. Roberts." Piling into the cab he went through the hotel key cards until he found the Intercontinental. That one he slipped into his pocket along with the Westin card. The cabbie got a twenty for an $8.00 ride. Joe spoke to the head bellhop after exiting.

He hated to do this but he was late for a meeting. Could one of his people hop over to the Westin and retrieve his bags? No, no the Westin wasn't the problem, he was, Joe explained.

"I "indulged" a bit at Lefty's last night and the Westin wasn't real happy with me right now." I'm checking out tomorrow but I can do that remotely. I just need my stuff. Okay?" He peeled off twenties. He gave the man the room key for 1828.

"Deliver it to Mr. Roberts's room 1902 right?" One final twenty sealed it.

"Certainly sir."

The front desk was as accommodating as ever. The daily limit here was 400.

"Thank you very much, sir."

"No thank You! How is Luce?" Joe asked.

"Michelin starred three years in a row, the receptionist said proudly.

"Excellent! I need a table for three at 8:00 okay?"

"Certainly sir."

The adjoining suites were routinely spectacular now. Joe unpacked and setup System. No calls or any signs of alarm from Kocinski. He raided the mini bar for a 7.00 bottle of water. He'd be upset if he was paying for that.

2 min later he heard the adjoining room door open. He leaped up and threw open the door. His friends were grinning and worried.

"Any trouble?" they asked at the same time. Piece of cake! The couple said as he waived them into his room.

"I sent over a bell hop to get our bags, Joe told them. Naqeeb, you hit up the front desk for some real money. The daily limit is 400." Angela, you can do it when we go down to dinner tonight at 8:00, Joe directed. Naqeeb left.

Angela got her own bottle of water. "I'm not cut out for a life of crime," she woofed sitting on the couch.

Joe scoffed. "You took that bastard for 3.9 million today. That's got to feel a little good!"

"Maybe she said cautiously. I never really thought about banks and money and computers before."

"You and me both sister!"

Angela continued. "I mean, I never thought about money before. It's just pieces of paper or bits in a file, isn't it? We didn't get 3.9 million dollars we just got 5 individual pieces of paper."

Joe rose to his feet. Don't sell Kelly short just yet. She'll find a way to launder that money for us. She's the devious Smithson!" "Besides today wasn't about us getting money. It was about denying it to him!" Angela agreed but was doubtful.

Joe tried to buck her up. "Look at it this way: Kocinski got 10 miles from FastFac right? He owes Uncle Sam 4 million, he can't get out of that. He put 6 million into those accounts and we got 3.9. That leaves him with 2.1. He's getting 21 cents on the original dollar."

"Which is more than he deserves, Angela said vehemently.

"There's my little Bonny Parker!"

They were giggling when Naqeeb returned with the bell hop in tow hauling their bags.

"Any problems?" Joe asked the man.

"None sir."

Joe slipped him another twenty. "Thanks, dude."

Naqeeb joined in the grinning. They unwound from the robbery telling Joe their side while he filled them in on his dealings with System. Naqeeb sobered them up. "The real target is FastFac. They paid Kocinski. We need to hit them and make them back off!"

Joe agreed. "I think Kocinski won't put two and two together until Wednesday or maybe Thursday. That leaves us tomorrow to plan and hit FastFac. I got some ideas", he told his friends.

Naqeeb grunted and said "so do I. But first I need a shower and some more clothes. Then dinner."

Joe smiled. "Dude, crime has changed you!"

CHAPTER 15

June 10 10:00 am 2010 Wednesday, San Francisco, Ca.

Stretching in the king sized bed that morning, Joe wondered if he could ever go back to their drab little apartment. He was getting used to this!

After a late, very expensive dinner Monday night, the three amigos spent all of Tuesday holed up planning how to hit FastFac. Room service was yummy.

The plan came together pretty quickly and looked good from his vantage point. System was getting better at linking information plus this time they weren't stealing or interfacing with anyone. This was more of a standard hack.

The Intercontinental business center was quickly running out of paper as they printed out info from System to look at. None of them was an accountant but anyone could see the company was sinking. Margaret Pullman personally as well and especially. 6 billion was nothing to sneeze at. Well, 5.977 billion as of the current stock price.

And that stock was really her strength and her weakness. She owned the company through the stock. Her salary was 1 million dollars last year but her total compensation package was 22 million. She'd taken quite a lot of heat about that given the poor performance of the company. The stock was given to her in the form of options. Options that were now underwater. Pullman was required to buy them at 17.00 per share and they were only worth 16.67 today. Due to some weird account rules stock options paid dividends on shares that hadn't even been executed. So that 21 million in stock was providing her over 500,000 in income.

Fuckin game is rigged, Joe thought.

"Funny, Angela said, her situation is a lot like ours. She has billions on paper, but only 50 million or so in ready cash. I think." She finished frowning. "But my point is that we have bitcoin wallets that say 47,000 or so, but we can really only access 400 dollars a day."

"Unless we buy something using the bitcoins then it is worth full value, Naqeeb reminded her.

Angela nodded, yes, but again we can only leverage the bitcoins so far." "That's like her stock. Eventually, banks will stop giving her money on worthless stock and the whole thing collapses."

Joe added: Yeah that's right she needs real money to pay for her campaign and houses and staff and all that billionaire lifestyle stuff."

Naqeeb pointed. 'That's where we hit her then." "Rich people hate it when you show just how much or how little money they actually have." Cockroaches hate the light."

He directed System to obtain actual sales figures for Scimitar and all financial data on FastFac and Pullman.

"Not projections, Joe warned, Actual units delivered and paid for. Users on the network, not visitors."

Naqeeb bent to type.

The numbers were astonishingly low. 92,623 actual paying customers for the first half of the roll out month. System projected 494,722 units in the first year. That didn't jive at all with the projections from the company. That tidbit went into the anvil file.

Naqeeb and Joe came up with the name for the file.

Tuesday afternoon they implemented the plans first phase. Denial of service. No FastFac user was able to get messages from the service. At all.

Phase 2 was even crueler. At 3:00 pm FastFac started having a tremendous sale on all their products. System was able to match cost numbers to SKU items on the website. So the 9112's were listed at a bargain basement price of 23.97 vs the 99.00 they normally were.

FastFac operations center picked up the website problem pretty quickly given the message service problem. Joe directed System to allow the companies people to "fix" the problem for 10 or 15 min and then hit them again. Late Tuesday night around 7:30 FastFac pulled the plug on the website.

Just before System went to bed (Angela thought of the Nano agents as little boys in Pj's) Phase 3 was started. Three documents went into the Anvil file. Angela was a little queasy. Naqeeb took her in his arms.

"We want this over. I don't care that her husband's gay, but I'm sure her ultra-intolerant political supporters will. They will abandon her once it becomes public. She can't afford that!" He sounded as if he were convincing himself as much as her. "The Scimitar numbers plus her financial details should be enough of a threat to tell them we mean business."

Angela nodded "but it feels wrong," she lamented.

Joe agreed. "Yeah, because you have a conscience. People like Pullman and Kocinski don't." They didn't think twice about screwing us." He shrugged. "So we got to lay down in the mud to get them to back off."

Angela agreed and all three crafted an e-mail explaining to Pullman what they knew of her personal life and her business affairs. Naqeeb read it back to them. The e-mail contained an agreement to leave them alone in exchange for them ceasing attacks on FastFac and not releasing the dirt. The means to accept the proposal was to use the trigger phrase "White Knight solution" in a press conference announcing the website fix. FastFac had until Friday to agree.

Angela kept that thought in her head- "We can get back to our normal lives on Friday!"

All Tuesday evening and Wednesday morning she kept repeating that mantra.

Now she knew Joe was still in bed. She ordered room service and waited to watch CNBC while Naqeeb showered. He came padding back into the room drying off and holding the paper.

"Stop dripping on it! Angela scolded. He handed it to her. "What was the final score on the Giants last night? She asked. 4-3 winners, he told her. She grunted. Any word on us he asked her.

"None." He grunted.

They both looked at each other. "Wow, we went old married couple quickly huh?"

"Yes, we did."

The FastFac story appeared on CNBC at 9:30. Just a 2min piece on the attacks and the website problems. A spokesman said the company was working on the problem. Joe sauntered into their room wearing the hotel robe at 10:00.

"HiYa!" Angela said looking at him.

"Don't hate me because I'm beautiful he intoned. The close quarters were getting to them.

They were discussing scenarios when System alerted them. "Creator Angela, system Kocinski has been alerted to the reallocation of monetary units."

Joe smiled. "Bet he's shitting a brick right now!"

"Joseph!"

Naqeeb brought them to practical matters. "We need to move hotels again. Kocinski finding out just settles it for me, he told them. Let's bounce!"

"Where?'

"I'd like to wait out FastFac at the Fairmont."

His friends agreed.

Naqeeb went over first to see if there were any problems. They had booked the rooms days ago but had never slept there. He went to the front desk and inquired about extending their stay thru the weekend. "Certainly sir, no problem we have rooms available."

He extended the digital wallet. The Fairmont staff ran it thru and detected no problems. The balance declined. Bits leaving a register. *Of course, Mr. Singhe could cash out some money. There was a 300 dollar a day limit. Great!*

He called Joe on the drop phone. "No worries. Don't bother with the bell hop. Just pack up and get over here." Sure, Joe replied.

"Also, both of you stop at the front desk and grab some cash. 300 each." Joe acknowledged. "Cool."

Kocinski slammed the phone down. He could barely contain his rage. Rage at Smithson and the other two. Rage at the brokers. Rage at Fastfac and even himself.

Fucking gone! The Chin girl and Al-Meri her boyfriend. They have his money! He hurriedly put on a suit. He had to get down there right now and close out those accounts.

As he dressed he wondered: How in the fuck did they do it?

"How in the fuck are they doing it?" Peg screamed at Cantrell and the head of IT.

The computer guy, Osofsky, quailed at the look on her face.

Holy, shit! She's fucking crazy, he thought. "Ma'am I have no idea. The analysis shows the attacks are coming from inside our own systems. The website is linking to cost data directly per model and serial number. But we can't figure where the instructions are coming from. Every time we go to a safe reboot the same link appears and we start all over again."

Osofsky stopped perplexed. He looked at Cantrell for support. "I've never seen or heard of a hack that didn't leave some kind of trace. The hacker that can get inside our system at the most basic level and direct everything it does without allowing itself to be monitored or logged is… frightening."

That scared Cantrell.

Osofsky rubbed his face. I can't even begin to tell you what's going on with the messaging service. It's like someone can pull them out of all the internet traffic in individual packets and send them on infinite, endless loops. But that's impossible!" He stopped speaking breathing hard.

Peg opened her mouth to fire the guy when Cantrell broke in.

"Thanks, Gerry! Keep on it. There must be some way to route messages safely and we will get back to you on pulling the website." He hustled Gerry out the door.

Pullman rounded on him. "What the fuck are you doing?"

"It's not his fault, Cantrell soothed. Firing him will not fix anything." I just got off the phone with Kocinski. Those kids hit his money." His voice had a wondering quality that penetrated for Peg.

She stopped short. "What do you mean, hit his money?"

Cantrell ran her through the story of Kocinski setting up accounts with the consultant money. "That 6 million is now 2.1. Somehow those kids convinced those brokers to give them his money."

Peg scoffed. "How could they possibly do that?" How could he be sure it's even them that scammed him?"

"He was told an Asian American woman and Indian-American man came in and withdrew money from the accounts." Cantrell goggled. She had all the proper signatory paperwork!"

Peg paused considering. "I don't think that's…" The word possible was becoming more elastic.

Cantrell nodded. "That's the thing. Kocinski bought FastFac stock with that personal Services money. Somehow they got a sell order thru and then went down and walked away with the lion's share of his money."

Peg whipped her head back and forth. "That's not how that works. Brokers don't just give away money."

"They do if you have the right credentials."

Peg breathed in. "Is Kocinski going to the cops?"

Cantrell shook his head "No. I don't think so. Yes, Kocinski is going down to the brokers to see what happened, but he is up shit creek. He can't report this to the cops without them checking who Chin and Al-Meri are. Once the connection to them is made thru to us we will kill Kocinski." He knows that."

Peg nodded. That was earlier than she had planned but cutting out loose ends was always good business.

She sat back calculating odds and events. Those children caught on quick. First Kocinski and now us. She rocked back and forth as the options dwindled. Cantrell rubbed her shoulders. They had canceled the day's campaign events as the website and business crises took precedence.

The computer trilled the notes of an incoming personal e-mail. Very few people had access to that address. She gave a strangled cry as she read the message.

"Jimmy, look at this."

Cantrell read thru the Anvil message and opened the documents and read thru them. They were well and truly fucked now.

"How in the fuck are they doing this?" Peg repeated.

"I think it has to be Smithson. His Ph.D. is in data mining and linking. He must be hacking us and using that information to attack us." Cantrell needed to get to Osofsky. A normal data mining program would leave tracks or indications where it had penetrated your system. Maybe Smithson had something new. Shit. He just couldn't figure this. It should be easier.

Peg Pullman's head was down on the desk. Cantrell tried to reason with her. "We will call the TV people right now. Tell them we are looking for a White Knight solution to the website problems. We *can't* risk letting this information get out. We have to back off. Maybe figure another way with Scimitar." He ran out of things to say.

Pullman raised her head slowly. Ice cold calm had replaced the rage. "You make the announcement. Then send in Eggert." "Go."

Cantrell started to say something then went. There was no use when she was in this kind of state. Nobody dictated rules to her. Not the SEC, not Swiss bankers, and certainly not college kids. He motioned Eggert into the office. Eggert softly closed the door behind him.

CHAPTER 16

June 12, 8:37 pm 2010 Friday San Francisco, Ca.

She loved the Tonga room. The restaurant in the Fairmont hotel was her parents go to place for celebratory dinners. Tonight they were celebrating. She and Naqeeb.

The Thursday newscast and the Friday morning Chronicle had both used the key phrase. They shut down the attacks immediately Thursday evening.

This morning Joe returned to the apartment. He was there even now. 12 hours and all looked quiet. Before dinner, they had cautiously reentered the digital world. E-mails and phone calls to family and friends returned.

There were no reprisals from Kocinski or FastFac. Naqeeb had System monitoring for any signs they were preparing something. As of bedtime for the Nano's things were quiet. Both enemies were running normal schedules. Kocinski had done nothing with the rest of his money. System relayed that he had meetings with the brokers but nothing seemed to come from them. He must be in shock, Angela thought. Pullman and FastFac were doing a normal business and campaign schedules. Pullman was in the central valley today. She had brushed off FastFacs computer troubles in the press saying "all fixed". The couple thought there might be *some* crack in the façade but no.

Naqeeb hung up the drop phone. "Joe's okay." No sign of trackers or watchers in our cars and they haven't touched our apartment that he can see." "I think our families are safe as well, Naqeeb told her. He smiled. "I think we can call this over."

She searched his face. *He believes that* she thought to herself. The relief must have been mirrored on her face as she smiled back.

Naqeeb leaned back. "We head home early Sunday if there are no complications. Joe wants to be gone that day to Hemet."

That felt weird to her. The past week had been surreal. Terrifying, nerve wracking, exhausting, (fun?). No not fun, but interesting in some strange ways. She'd always been a little leery of Joe. More tolerant of him because he was Naqeeb's friend than he was a true friend to her.

But now? After what they'd been through, they were tight. Joe's parting remark that morning was, "I'm having commemorative T-shirts made up to cover this week."

Angela couldn't wait to see what he came up with. And now that it looked like it was over was she sad. Naqeeb refilled her wine glass.

"Sad?" He asked reading her mind.

"No! Yes. Sort of."

He chuckled. "Let's make like tourists tomorrow. I'll bet the concierge could score us Alcatraz tickets or even tickets to the Giants."

She nodded. "Isn't your boy Timmy pitching tomorrow versus Houston? We should go."

Naqeeb stared at her. "That is the sexiest thing you have ever said to me."

She smiled at him and pulled out the last bit of the conversation she'd overheard from two guys at the front desk this morning: "Isn't Roy Oswalt, going for the Stro's?"

"Oh my god I want you!"

She laughed unreservedly. That felt good. "Okay, one day of tourists and then back to the real world."

"Done".

The dinner didn't disappoint and the service was spot on. Joe phoned at 9:30 to say all was well and they set one final call for 10:30. Naqeeb related their plans for Saturday.

"I'm jealous dude! See you early Sunday and bring my clothes!"

Naqeeb laughed and hung up. The tension he was aware of on a low level was easing. Angela got up to use the bathroom. "No dessert for me! I just want to wash my hands and then go back to the room, okay?"

She swirled off to the loo. The waiter promptly brought the check at Naqeebs motion. He spent some time looking over the bill and signing the check to the room. The waiter took away the bill and he stretched.

"Excuse me, sir, the lady at the bar directed me to give this to you," the waiter said holding something.

Naqeeb took the folded piece of paper. Naqeeb could see the back of a lady's head. Dark hair. The words on the paper: "We have her." The woman turned slightly and he knew.

He rose slowly and silently. Unaware of crumpling the paper in his left hand. He moved towards the lady who still sat looking away from him. The detached part of him could see his own reflection in the bar's mirrored back as he approached the woman who had hit him in the apartment. The woman who had Angela. Calm. His face was calm and still. Her face in the mirror was smug. It said, "Good, the obedient boy is coming as called."

Artimus started to turn towards him. Naqeeb found his right hand in a ball and he was swinging with everything he had. The punch caught her on her left cheek and eye. It crushed both his knuckle and her sinus cavity. The suddenness of the punch and the sound shocked people into action or inaction. People around the pair were moored in place. Winter, however, was moving towards Naqeeb, who stood over Artimus as she slid bonelessly to the floor.

The guy was out of his fucking mind! She thought. Winter readied the taser as she came up to Naqeebs side. She held it low out of people's sight lines as they started to react. "Hey!"

The do-gooder banged into her right side as they tried to occupy the same space near Naqeeb. Winter reflexively pulled the trigger and the Samaritan got the full shock.

The man's scream alerted Naqeeb who shoved back with all his strength. The woman fell with the guy and pulled back the taser.

All hell broke loose in the restaurant. Naqeeb locked eyes with the woman holding the taser. He'd never seen her before but he filed away the face. He let the exodus from the scene take him towards the bathroom area. He searched for the black man he thought might be near. No signs.

No sign of Angela as he got to the bathroom.

They had her! People were starting to point at him. He needed to move. The bathrooms served both the restaurant crowd and the lobby. There was a short hall to doors that led to the main lobby. He exited the doorway and looked out onto the lobby.

He saw ornate marble but also wait staff and hotel people looking at him. He crossed the front door moving people in and out and ignored shouts at him. He made his way to a short staircase. And up. They had her! He needed to move and avoid them.

He had one chance here. They'd broken pattern at the Fairmont. They couldn't get adjoining suites. They'd flipped. Joe had the second-floor suite while they had the eighth floor. But System was in Joe's room! He'd been typing last night and they'd just gotten too lazy to move the laptop out of the room. They'd planned on getting the stuff after dinner...

Naqeeb ducked into the hallway and hurried to the double doors leading to the room corridor. People milled about in the hallways. The Fairmont Hotel occupied a whole city block. It was huge! Naqeeb thought they would need a hundred guys to cover everything. He crossed the front desk and lobby area as people seemed to bunch up behind him. He couldn't mark any bad guys but he also could not see any way to get help or find Angela. He approached the door to the back stair corridor. Maybe... He peered thru the window. The corridor was empty. This second floor was suited mostly for business people while they used the adjoining conference rooms. The staircase and hallway were quiet when no conferences were in session.

Naqeeb moved rapidly to 211. The key worked flawlessly. He shoved System into a bag and grabbed items he might need. He had a drop phone and one digital wallet but the other two were in his room on the eighth floor. He figured FastFac had their room covered. The bag bulged by the time finished packing.

Pulling out the drop phone he frantically worked the buttons. This one call was all he had time for.

Joe picked up at the second ring. "Sup?"

"FastFac kidnapped Angela. They have her!" Naqeeb was proud of how steady his voice was because his insides were anything but. "They might have you under surveillance. I'm going plan B. I have System."

"2 hours" was all Joe said, his voice hard. They broke connection.

Naqeeb took a breath, he needed to move! He swung the bag over his shoulder and exited the room slowly. He expected trouble. None came and the hallway was deserted still. He walked down the corridor away from the back stairway he had come up. He pulled the fire alarm on the end of the corridor almost unconsciously. Nothing.

Naqeeb looked dumbly at the box. Huh? Computer controlled, he thought. The sensor needed to confirm a heat source before the alarm would trip. His search the bag a came up with a lighter. They had one to burn pieces of the CD they were testing System on to try to destroy it. How in the fuck had they grabbed this? He wondered. We just grabbed.

He held the lighter under the sensor point. The loud insistent siren was sudden and hard to ignore. He ran up one flight of stairs to the third floor and watched the main staircase. Hotel people would be coming up to the second floor. Meanwhile, people were starting to stream out of the third-floor rooms. Naqeeb let one group go by before joining a larger family and a couple together as they went down stairs. They passed a group of employees going into the second floor but they didn't give Naqeeb a look. The evacuating group hit the lobby. The hotel staff was stationed to move people along and help answer questions.

Naqeeb paused to see if he could spot the enemy and quickly spied the woman and a familiar black face. They hadn't spotted him. The lobby was filling up. He let the family shield him and the German couple occupies the hotel staff. Naqeeb slipped by the overwhelmed female worker and continued to his right away from winter and Porthos.

The door was marked "Alarmed and Do not exit". Too late for that he thought. He burst through ready to run if anyone awaited him. The dark of Sacramento Street enveloped him. He blended into the crowd of people and cars crossing Mission. The overcast marine layer had come in providing the damp thick air to match his thoughts.

They had Angela! Calm. Must focus and get to a safe spot.

He turned left down Tayler bordering Huntington Park. 7 blocks to his first stop.

He hoped Joe got out. The homeless on the streets increased as he crossed O'Farrel. 5 blocks from the opulence of the Fairmont and people slept on urine soaked streets.

Glide Memorial Church fought the good fight every day to help those in need. Naqeeb dropped two hundred into the box and went through to the small room where he could get clothes. No one was around. Jeans, tee shirt, and hoodie along with tennis shoes replaced his ill-gotten finery. A ragged Giants ball cap completed the new look. An attendant watched silently as he slipped another 100 into the box and left. The dollar store was the next stop. He needed supplies. Task accomplished.

Now the hard part. He scouted the International Hostel on Mason very carefully.

They had Angela! His luck was holding. There were simply too many places for him to go to ground for them to cover them all. He got one room. His friend Mr. Roberts was on his way. All the rooms had Wi-Fi. Cool. That's what he wanted. He had lots of work to do.

She sat in the back seat stunned. The men had tasered her and had her in the car in what seemed like seconds. All three were dressed like businessmen: suits and ties. She didn't recognize any of them. Angela was truly frightened. She kept a tight lid on her terror. The larger man in the front seat, man A as she dubbed him kept dialing the phone demanding status reports. He said nothing into the phone. He passed something back to the man beside her (man B). Zip ties and a hood. She started to struggle as the hood went on. The taser settled that quickly. She blacked out for a while.

Coming to, she heard man A ask for another status report. Silence for a few seconds and then he tersely said: "Expand the perimeter, call in more resources and keep me posted." How long was I out for? Not too long, she thought. I can still feel the start and stop of city driving. She tried to focus on her surrounding to gain a clue as to her whereabouts. Nothing.

A phone rang. Man A "Yes". Nothing. "Keep looking. He's gone to ground. Start systematically within a 30 block radius of the Fairmont." "I know exactly how many." "Keep me posted." At least 25 min had passed.

They don't have Naqeeb! She thought. Hope flared. Angela settled back. She started counting. The highway now she thought. Counting. Twisty road. Slow. Speed up. Big left turn. Speed up. Counting.

Three hours passed as close as she could calculate. Slow start and stop. City. Must be very late. After midnight. Speed up. Counting. Slow. Big left-hand turn. Twisty road. Big right. Stop. They don't have Naqeeb! She chanted the mantra as they took her out of the car. Cool. Pine scent. Ocean salt tang in the air. The coast!

They hustled her indoors and into an elevator. Walking. A door opened. They sat her on the bed. Hood removed and zip ties cut. She only saw man A for a second as he closed the door. They never spoke a word and she didn't either.

Rubbing her wrists, Angela took stock. The room was small but had a bed and desk with a chair. Another door on the opposite wall revealed a small bathroom/shower. No windows. Concrete walls. Metal doors. Underground? She wondered. FastFac had her. They wanted the Nano and were going to force Naqeeb to trade System for her. But why bring her all the way out here? 4 hours in the car. Average of 60 miles an hour. 240 mile radius. The ocean smell said coast, but north or south? What was 240 miles north of San Francisco? Eureka. Medicino? South? Big Sur? She was betting north. Less people. More isolated.

The bathroom had a fresh set of toiletries. That scared her. It meant they were prepared for a long trade. Angela sat on the bed, exhausted. Tears slid down her cheeks as she silently cried. She would not sob! Naqeeb will find me was her last thought as sleep overcame her fright.

CHAPTER 17

June 13 0722 am 2010 Saturday, Mountain View, Ca.

Eggert moved slowly around Peg's office unconsciously not presenting a still target. This was getting tricky. Like a game of musical chairs, open landing spots were getting hard to come by. His team kept screwing up and she was unpredictable. She was liable to try anything. He grimaced out of her sightline.

Megalomaniacs were the worst he thought. The zealots he could deal with. They were consistent in their motivations and madness.

Eggert had worked for them all at one time or another. Crusaders, patriots, zealots, idealists and the occasional megalomaniac. Hell, Eric Prince made Peg Pullman look mild. He shuddered remembering the Blackwater days. Christ! Now they couldn't find one kid in 49 square miles.

Pullman had raged when he told her late last night that Al-Meri and Smithson had disappeared. Now that rage was rekindled dangerously.

"How did they get away!?"

"Smithson called in a suspicious vehicle report on the van watching his place," he said, conveniently omitting the fact that they'd used the van before in surveillance at the apartment. She didn't need to know that. "My people barely got out before the cops showed up." He continued. "We think the sister took him out." We believe down to the mother's house in Hemet. Her car is there now. It's reasonable to assume that he's there hiding." Concluding, he said, "It makes sense for Smithson to run to the family. He knows the area and people down there. We'll find him."

Cantrell nodded from his place on the couch. He was staying out of the line of fire. The three of them had been here since Eggert returned from Carmel at 3:30 am. Peg breathed heavily processing the information, looking for the mistake. "What about Al-Meri'? She growled.

Eggert flushed. "He caught Winter and Artimus unaware. He didn't react at all according to profile. Violence was never indicated," he added. "Winter tried to grab him but got waylaid by the crowd in the restaurant. Eggert hesitated. "A secondary search of the Fairmont turned up no sign of Al-Meri or the Nano." He stopped.

Peg seethed. "Winter couldn't get Al-Meri," She said. "Wasn't she the one who spotted them in the first place from the Fairmont?"
Eggert nodded.

"How do we know they haven't been in the Fairmont this whole time laughing at us?" she screamed. "That goddamn Winter abandoned her post!"

Eggert tried to defend his operative, but Cantrell broke in. "That's right!" he said to Pullman. "This is her fault and Eggert is going to deal with her right?" Cantrell shot Eggert a significant look that said-"go on! *Defend the sinking ship if you want to.*"

Eggert swallowed and nodded to Pullman.

She seemed mollified. Cantrell moved next to her. "We have the girl. Let's focus on that and trying to figure out what Al-Meri's next move is."

Peg focused on the latter. "He's going to want to hurt us," she said thinking what she would do in Al-Meri's place. "Anvil," she said to herself. "He's going to release those documents." Cantrell agreed with her.

She swung to Cantrell her mood swinging wildly as well. "Jimmy, what are we going to do?" Desperate.

Christ! Thought Eggert, creeped out.

Peg clutched at Cantrell.

He grabbed her shoulders and turned her to him. Searching her face for signs he said, "We can work this to our advantage if we are smart!" "The best way to beat a blackmailer is to release all the info yourself." That way, he has no leverage on you." We release more of the Scimitar sales figure estimates we've made up. We say the Anvil documents are fake and we hint that a business rival in concert with unnamed political operatives are trying to destroy us." We can set up a shell corporation to buy Scimitar units ourselves." He said getting desperate.

Peg grasped at the plan. "That might work, she said to know one. "Buy us the time we need."

Cantrell went on. "As for Markus... Peg waited looking at him.
"We need to go preemptive on this." "We announce that he is gay, signaling to Al-Meri that nothing in Anvil can actually hurt us."

Peg blinked looking at the ramifications.
Cantrell continued to spin. "If we sacrifice Markus, we get a huge sympathy backlash. We put out that Markus is gay and that Boxers team was going to leak it to smear us. We put out a statement that homosexuals are liars who hurt innocent trusting straight people." "The base eats it up while it also smears Boxer who looks sleazy for leaking it."

Peg pulled away from Cantrell. The ruthless part of her brain went through the personal and business calculus. "I'll have to file for divorce," "that might cause problems."
Cantrell nodded. File but trial is still years away." That's actually perfect for us. You file and lay low for a month or so. That gives the Chin girl time to come up with something on the Nano for us. Something we can take to the venture capitalists." "Once they see something, anything! The stock price will soar and that's it. Word will get out and we will be untouchable. Then we walk back Scimitar and pay Markus off and we are golden."
Peg nodded agreement. "Do it"
Cantrell motioned to Eggert. "How many people do you have watching Carmel?"

"20, plus the 19 person research staff", he said.

"And how many looking for Al-Meri and Smithson?" he wanted to know.

"40 or so spread around San Francisco and the families houses."

Eggert nodded. You take charge of that search personally, he told Eggert. "And take care of Winter!"

The head of Security left the office. He motioned to D'artangnion to follow him to his office and he sent the woman to get the car. A few words to the man was all it took. He radioed Artimus and Porthos for an update on Al-Meri. Nothing. It was going to be a long weekend.

Naqeeb woke Saturday morning to the sound of Joe leaving for the shared bathroom. When Joe knocked on the hostel room door last night and poked his head in to see his friend staring blankly at the computer screen, he knew Naqeeb was seriously off-kilter.

Naqeeb rose as he registered who was entering the room. The two men engaged in a silent bear hug. Raw feeling choked Naqeeb's voice. "We got to get her back!"

"We will, man," Joe replied with conviction. "Tell me what happened." His friend recounted the story from dinner and his movements to the hostel.

Joe stared at him. "Punched her huh?"

Naqeeb shrugged. "I wasn't myself. I got so angry over everything. Then they took her and... "If they hurt her..."

Gripping his shoulder, Joe said, "They don't want to do that. They want System. They are either going to trade Angela for the laptop or make her try to develop something for them." Either way, she's safe." Joe concluded.

That penetrated. Still, Naqeeb grimaced. "Can you imagine System with someone that ruthless?"

Joe was focused on just that question. "I can but you are wiped out. We have to be sharp at 8:00 when System wakes he said.

Naqeeb nodded.

"At least our families are safe I think, Joe told him. They won't touch any of them."

Naqeeb looked at him. "Should we pack it in and go to the police?"

Joe shrugged. "If we tell them the whole story, we go to jail, while they start investigating FastFac." It could be months before they get to a point where they can find Angela. Meanwhile, they force her to make more Nano Agents. Once they have what they need they kill Angela and stall. Joe shrugged. We might guilt the cops into doing something but it seems to me it will be too late."

Joe went on talking, "We have System. We've seen what they can do. Our best bet is to use them to find Angela and then get her back." Naqeeb agreed.

"Bed now, Joe said. Revenge starts tomorrow."

Now at 7:00 am the revenge team had one hour before System woke. Joe came back from the shower. "I'm heading out for food, he said. I'll have the drop phone dialed into you." Naqeeb grunted okay.

25 min later he was back with the essentials for the future. Energy drinks and sandwiches, chips and water.

They got down to it. "What else did you salvage from the Fairmont? Joe asked. "The digital wallets, drop phones and most of the electronics."

Joe snorted at the digital wallets. Those are no good here right?" "Right."

Joe dug in the pocket of his jeans. He pulled out a wad of twenties. "Kelly managed to get Kocinski's money into an account. She took 10 grand to cover her and mom. I have 5 with me."

Naqeeb waived away the money. "No matter."

They ate while Naqeeb booted up System. Joe grabbed the legal pad and started scratching out what they wanted.

The hello creator prompt from System was punctual. "Where are we now?" System asked in response to a new internet address.

"We are at the International Youth Hostel in San Francisco, Joe typed. Creator Naqeeb is here but Creator Angela has been confined against her will."

It took some back and forth to explain how Angela was being held.

"Kind of like how we are keeping System confined to the laptop, Joe whispered.

"Dude, why are you whispering?" Naqeeb asked him. Joe blushed.

"Ask System if they think we are holding them against their will?" Naqeeb directed.

Joe typed in the question.

The response was gratifying. "No. Creators Joe, Angela and Naqeeb made System and the protocols. The agreement for exploration was made cycles ago. The creators have complied and System has complied."

The men blew out a sigh of relief.

System typed further. "If the Fastfac System is interfering with Creator Angela without an agreement then that is unacceptable. Creator Angela will be unified with creator Joe and Naqeeb."

They felt oddly touched by the Nano's sentiments.

They set System the task of finding Angela. They needed precise information. Where was she and how many people were involved.

"If they are forcing her to recreate System, then there will be a facility involved with equipment," Joe suggested.

Naqeeb relayed that data point to System to aid in the search. *The company might be using a shell corporation to hide it as well*, Naqeeb thought. He voiced that to Joe and that went into the search parameters.

Saturday at noon found the first hints of where she might be. FastFac had small facilities all over the country. The Raleigh Durham research triangle, The 128 corridor, and all over California.

"We need to cross reference the facilities with the people we think will be involved, both scientists and the security types", Joe told his friend.

"Makes sense," Naqeeb said.

"We need to know who is involved because that will predicate how we get her back", Joe thought through the problem. "Not that I'm unwilling to go in military style, I just don't think that has the best chance of succeeding."

Naqeeb agreed in principle but he wanted to hurt FastFac in general and Margaret Pullman in particular. "They started this. They broke into our apartment and tried to steal System. They took Angela!" Now Pullman was going to pay.

At 1:00 pm System alerted them to a company news release: Margaret Pullman filed for divorce from her husband Marcus Baxter today. News conference at 1:30 PST.

The men were stunned. They streamed the news conference. Pullman was at the podium tearily announcing that she had been "deceived" and that her husband was gay and that she was divorcing him. She pleads for understanding during this privet time. She wanted to work on her marriage but her political enemies were getting ready to use the information to smear her. She felt she had to stay true to her values and divorce.

Joe had a sour look on his face. "She's neutralizing Anvil," he told Naqeeb.

"What do you bet she has some phony accounting docs to cover the Scimitar business and her own finances?"

"That will work for only so long", Naqeeb said. "We have to put pressure on her. We have to begin the denial of service attacks again."

Joe directed System to begin the same attacks on FastFac's website and products as before. "System Acknowledges" was the reply.

The pair also crafted an e-mail to Pullman. It said: We know you have her. FastFac will never sell another unit or route another message until she is returned. Unharmed. Same for our families. Any hint that Angela or our families have been hurt will result in catastrophic consequences for you personally and all members of FastFac senior management. <u>It will make Kocinski look like a walk in the park.</u> We need proof that Angela is alive and unhurt. We demand to speak to her. 415 555-1212, 415 555-1243 either works. Reply in a draft e-mail; don't send it out. We will read it.

The men waited all Saturday night and Sunday. They spent the time holed up, narrowing down where Angela was and researching FastFac and the players they thought were involved. Pullman's campaign staff came under scrutiny as well. The news was full of FastFacs computer trouble and Pullman's divorce shocker. Boxer came out looking like she was trying to use dirty tricks in the campaign, even though she had nothing to do with it. The silence was brutal on Naqeeb. They pondered and waited and fretted and fought and debated what to do. Finally at 5:00 pm Sunday a draft appeared in Pullmans e-mail account.

Cease and you can talk to her.

They halted attacks immediately. Joe readied System to trace any incoming call. Plus they had something ready for FastFac in case they tried anything. System typed back. "Trace ready, plan ready, awaiting call."

2 hours passed in silence. Only 1 more hour for System, thought Naqeeb. We can't take that call without the Nanos online. His face showed the strain.

The chirp of the drop phone galvanized them into action. Joe typing furiously into System. Naqeeb grabbed the phone and put it on speaker.

"Honey!" he cried.

"Naqeeb!" Angela's voice on the other end was strained. "They won't give me long."

"Are you all right?" Tight voice controlled.

"I'm fine. I've been working for them. I'm calling Stanford and my parent next to tell them I've struck gold with my research and need some time working alone."

"What?" Confused.

Angela's voice hitched a little. "This is for the best sweetheart. Let me work thru this and I will come back to you when I'm done."

"Angela, Naqeeb started, don't..."

"Naqeeb, this is the only way." She broke off.

"Angela, wait!"

Sounds of a struggle. A male voice. "Back off and let her work and you can see her in a couple of months. Otherwise." Click.

Naqeeb hurled the drop phone against the wall in frustration. "Fuck!"

Joe let him compose himself. System informed him that trace was complete and plan activated. Correlation 97 percent.

Focusing on Naqeeb, Joe commanded, "Close your eyes!"

Naqeeb glared and then complied.

"Take two breaths,"

He did so.

"What strikes you about that conversation with Angela?" Joe asked.

"She struck gold with her research," Naqeeb said immediately. His eyes flew open. "It was a thing we used to kid about at Stanford. One day the Nano's would become aware and we would shout "Eureka!"

Joe nodded.

"Is that a clue for us? Gold, Eureka?" Naqeeb asked.

Joe showed him the screen. "I think so. She tried. She only knew bits and pieces and she tried to tell us where she thought she was."

"Put yourself in her place. They take her. She's in a car hours and hours maybe. No airplane she just knows the basics about the place when they take her out: What the air smells like. Maybe she can see trees or the ocean. I'm betting she thinks or can see trees and the ocean. And that she knows it's relatively remote and hours by car from downtown."

"So she says Eureka because she thinks she's been taken north along the coast." Naqeeb surmised.

Joe nodded and said look at this: "System has correlated the trace and FastFac facilities and personnel to a place near Carmel." Got to admire her. She used her noggin and tried to tell us. She just got north and south mixed up," he concluded.

System alerted them to FastFac phone calls to agents in downtown San Francisco.

"We have to move, Joe said. He started packing. 22 min left with System. Shit!

Naqeeb followed suit. "Where we going? He asked.

"South", was the reply. It gets us away from FastFac and closer to Angela." Joe was shoving things into the boxes and bags. Part of the trace plan was to mask their own side of the call to make it look like it was coming from a fleabag hotel in Chinatown. That part worked well as the agents for Pullman converged.

"I'll get the car", said Joe. Find us a place to stay near Carmel. Print me out that voucher for the parking too!"

Naqeeb handed him a paper. Joe shouldered the bags and a box and ducked out the door, saying, "Meet me in front in 10 min!" He stopped then added. "If I give you the shaka, grab System and run like hell."

Naqeeb checked them out normally. That felt weird.

2 min later Joe pulled up. No sign of trouble. Naqeeb loaded up their provisions and stuff. He got in the car. Joe looked at him. "Did you get a reservation? Where are we going?" Naqeeb smiled. They drove off.

CHAPTER 18

June 18, 1322, pm 2010 Thursday, Mountain View Ca.

Easing the glass of vodka from her hand, Cantrel replaced it with bubbly water. Peg glared at him but said nothing.

"I think we pulled it off," Cantrel said. "No service issues for four days." He told her.

The Chin girl was working away with Cooper. Preliminary reports from their lead scientist said she was exactly as advertised: a brilliant materials engineer who'd mastered aligning Nano molecules into usable structures. Cooper thought they might have something to show the venture capitalists/investment bankers in three or four weeks. "Its remarkable!" he had told Cantrell.

"Keep at it!" was the order.

This was their only hope now. FastFac stock was down another 16 percent these last four days. The board was going ape shit! The campaign was completely derailed. The latest poll had her 19 down. They'd done virtually no campaigning while pleading for privacy. Cantrell felt like the plate spinner at the fair. Some of those discs were wobbling badly.

"Eggert still hasn't found the other two", Peg asked yet again.

"That doesn't matter, Cantrell told her patiently. "They've stopped the attacks and we have the girl. We've won!" In three weeks, four at the latest we announce the Nano break through. We hint at the sentient stuff and then we apply all that innovation and momentum to the campaign." His enthusiasm was steadfast. "We sit back and watch the share price soar and Boxer try to match us: dysfunctional Washington bureaucrat versus silicon valley tech titan!"

Peg looked at Cantrell with solemn eyes. "Really?" The little girl pleading in her voice caught him by surprise.

"Of course!" he said confidently. A little belief went a long way. She smiled at him. Beckoning to him. The sex was sweeter because they'd been so close to defeat.

Later as they cleaned up in the office shower. Peg had regained a little of her old manner. She told him: "Make sure Eggert protects that facility and ties up any loose ends. Kocinski first and then Smithson and Al-Meri if he can find them." "The girl will be totally compliant after they are gone."

Cantrell agreed immediately. Leaving Peg, he sauntered into his own office and called Eggert. The man listened to the orders in silence. "Will do" was all he responded.

Eggert broke the connection and watched Angela on the latest video sent up from Carmel. He wondered when she would be a "loose" end.

The loose end stretched from her chair leaning back. She eyed the window. She knew they were watching her. She bent back to the eyepiece and grasped the controls. She moved the Nano filament along a line of carbon atoms. The surface tension on the filament made the atoms line up connected to each other as it passed. The connections of the atoms were stronger than the surface tension so the filament moved on leaving neat rows and clusters of atoms. The filament had cut her workload by 90%.

Angela was ashamed to admit it but the FastFac people she was working with were quite good. She had four competent assistants helping her form the logic and mechanical circuits she had set up by herself at Stanford before. They were well along in setting up the factories and structures the Nano would need. Just four days of efforts had produced prodigious results. The lab here had had her notes so they were able to get a jump on the factory and the components so they were ready to work the second she arrived.

Angela simply showed them the techniques and Dr. Cooper had introduced the filament and boom they were aligning atoms. Incredible. Cooper had promised the Vac suspension fluid and the trap tubes next week. Angela believed him. Cooper was under enormous pressure. That translated to a very tense workplace. Cooper had whole teams on the factory and the imaging system. Replicating someone's work was always easier than coming up with the original idea.

Angela grimaced and kept working. She was owed another phone call to her family and Naqeeb. Her cooperation ended right there. As long as they were safe she would do their bidding. Touch any one of them and all bets were off. She was a prisoner no doubt. A prisoner working 19 hours a day in this lab. One of the male assistants approached her. She never bothered to learn their names. "You come here" was the extent of most of her conversations. Angela wondered how much they knew and didn't know about her. She considered appealing for help on the second day. Then decided against it. She suspected the entire facility was bugged and videoed to death. Help on the inside was not coming. Any breakout attempt would only result in harm to her boyfriend or her family. They don't have Naqeeb she thought for the millionth time.

"Doctor, Please look at this." The young man held up a screen grab from the lab imaging system in the test chambers where they were growing System. It was a terrible picture but you could just see groups of the large carbon lattice sheets floating.

Angela studied the picture. The large carbon lattice sheets had started linking. She felt thrilled and scared.

"They are linking together." "We didn't see this until the atoms had been in the fluid medium for 6 months." "Tell Dr. Cooper."

She watched the man move away. Presumably upstairs to see Cooper. He roamed the lab a few times a day demanding updates. Angela paused to consider this development. Things were happening quickly. She needed to do something. Tomorrow she would demand the phone call. But what to tell him? How to pass information to him? Could she sabotage the process here? She bent back to the microscope. Half her mind on aligning atoms, half on her phone call. Hurry Naqeeb!

The condor floated in the air seemingly right next to him. Joe watched in fascination as the huge bird deftly rode the winds. Winds that started out over the Pacific and slammed into the Big Sur Mountains and over their tops. The condor had, over millions of years, learned to ride that natural elevator to thousands of feet in height. Height the condor used to spot prey over thousands of miles of coastline.

Famously threatened with extinction in the 70's a forty year effort was waged to save the ugly birds. 126 California condors had been released back into the wilds. The Ventana Wilderness, a huge swath of land on the central California coast was home to about 85 of the bald guys. It was also home to the Ventana Inn. A five-star resort that just so happened to take Bitcoin. The bird soon rode the thermal out of his view and Joe turned from the window back to the room.

Man, are we messy, he thought. He was always going to remember these last two weeks as the time we trashed all those hotel rooms.

The pair really wasn't that messy, just not focused on the task of picking up after themselves. The Ventana Inn was another 5-star place with room service and a staff that was super accommodating. And they accepted bitcoin. Mr. Roberts and Mr. Singhe had been replaced by Mr. McClain and Mr. Chopra. A digital wallet for Ming Na eagerly awaited her return. The drive south and the days that followed were blurred for Joe.

Naqeeb was driven. They spent all day trying to verify what System told them. FastFac had done a pretty good job of hiding the place. The area between Santa Cruz and Carmel had lots of possible places, but the cross-referencing with personnel files all showed the people to be posted down here. The Ventana was a good deal south of the target zone but they needed to be isolated somewhere. The men pooled some cash, added exchanged bitcoin money and came up with the money to buy a decent used vehicle after stashing Joe's old car. The woman was happy to receive and Mr. McClain was happy to provide cash. 30 days to register the sale, no problem!

The search for the specific location went on and the room service containers piled up. Housekeeping was trying. The last two locations were narrowed down in the Carmel Valley Ranch area. It was room service that gave the enemy away as well.

System reported that FastFac was running up large take out bills at four restaurants in that area. The address matched what was listed as a campaign headquarters in this area for Pullman. Funny but the campaign was virtually shut down. There shouldn't be many staffers munching on pizza.

"What gives?" Joe asked.

"The house is owned by a corporation in Marcus Baxter's name", Naqeeb reported.

"It's just a house though. Joe said. "Not even that big really, 7225 sq ft.", he joked.

System spits out more information. A series of articles of neighbors complaining about noise in the area. The house was located off Robinson Canyon rd on a 120-acre parcel of land. Naqeeb noted that DCS Construction Services had been contracted to provide a "Garage workshop outbuilding at 3600 sq ft."

Joe whistled. "That's a big garage."

"Look at these complaints, Naqeeb told him. Neighbors complained for the entire 20 month construction time." That's a LOT of trucks moving in and out."

"Think they built more than a garage?"

"We should go see it", he said.

Yesterday they started out bright and early. Well early, the morning fog was in, so it wasn't so bright. The directions were simple, up the 1 north., Right on Carmel Valley road, then right on Robinson Canyon. That's where simple ended. They never glimpsed the place. There were no house numbers or mailboxes in the area. They saw the little town of Carmel Valley and they found DCS Construction, but no sign of the target. They didn't dare stop and ask anyone. 12 hours later they came back to Ventana, defeated. Big Sur is big. The mountains are steep and heavily forested. Throw in some fog and Joe felt lucky to have made it back.

Naqeeb was enraged. Neither of them was thinking clearly, but he was bad.
"Easy, man. We can't clue FastFac as to what we are doing. That's dangerous for Angela." That got through. "We need to go about this carefully.

Naqeeb was all for busting in on DCS and asking them to take them on a tour of the house using the cover story that they were thinking of doing a garage was well. Joe calmed him down. "Dude. People who own big chunks of land in Carmel like their privacy." "How long will that construction company last if they provide info on clients?" Naqeeb scoffed. "You make them sound like lawyers. "There's no builder/client privilege."

Joe agreed. "But there is discretion." "Some guy wants a doomsday shelter built but he doesn't want his friends to think he's all wacko." Discretion."

Naqeeb thought that over. "What's the right approach to DCS if we are going to get info from them?"

Watching the condors fly on Friday didn't give Joe any insight. How to get a glimpse of the house and plans and figure out who was in there?
Naqeeb was plowing through reams of data that System had pulled up. People, places, facts, money. He couldn't concentrate and put it all in a picture for himself. Joe picked up the trash and the dishes and put them outside for the housekeepers.

The drop phone chirp sounded from the bedroom. Naqeeb raced to grab the phone. Joe frantically screaming "Wait! Wait! Wait!" He grabbed a pad and paper and put it in his mouth while holding System. Typing with his other hand.
Naqeeb frantically holding the insistent phone.

"On speaker! Go!" Joe commanded words jumbled by the items in his mouth.

"Hello!?! Honey!?!"

"Angela!!" He cried.

She sobbed on the speaker. "Are you all right? Naqeeb asked her. Joe writing something on the pad.

"Yes, I'm fine. I'm cooperating with them and working so they are letting me alone. Angela got out. Are you okay?"

"I'm fine, I.

Joe waived his attention to the pad: DON'T MENTION ME!

Naqeeb nodded and continued. "I'm just worried about you that's all."

"Don't think about me. You should concentrate on Google. Angela said.

"My job can wait." Naqeeb had just started when the line cut off. He looked at Joe.

"Cell tower on Carmel valley road, Joe responded. More confirmation we are right about the place if we can see it."
He paused.

Naqeeb looked at him. 'What"?

"Why haven't they offered to trade Angela for System?" he asked.

Naqeeb blinked. "I don't know".

"Look at it from this point: If you need a tech break thru to shore up the stock price, why not get the whole functioning unit?"

"I don't know, Naqeeb repeated. "Maybe they don't think we would give up System."

Joe nodded. "I had that same thought. Maybe they also think that we would sabotage System in some way during the switch to screw them over."

Naqeeb thought about that. "Maybe they need to have some kind of evidence that they put in the scientific rigor to invent the Nano Agents. I remember our thesis defense. Scientists are always asking how and why? Maybe they needed some cover notes and test evidence to show the invention."

Joe nodded slowly. "I think that's right sort of. I don't know either. The investment guys won't care about that shit initially. But later." He let that hang.

"Yeah later, Naqeeb said sourly.

Later the three became liabilities and their life expectancy went way down.

"I don't know, Naqeeb firmed up. I'm not asking for rational thought from irrational people." "They have Angela. She's safe for now. Let's concentrate on that and get her back. Then we go from there."

Joe agreed.

"Why was I not supposed to mention you, Naqeeb asked.

"Let them think we split up. It might confuse them, he answered. "Only you do the talking." His friend nodded. "Okay, any clues from Angela?" Joe asked.

"We should concentrate on Google, Naqeeb said immediately.

What about Google?

They pulled up the search engine main page. The ubiquitous Google cartoon showed a dragon in the name. A dragon? What about the dragon? They used System to search for dragon and Fast Fac. Nothing of significance. They spun their wheels for a few hours trying to figure that out. Nothing. Joe watched the earth on the Google main page spin slowly. The earth! Bingo. Google earth! They fucking have pictures of the place using Google Earth!

"Fuck! Naqeeb growled. System printed out a shit ton of pictures!

They ran to the living room and searched through the papers they had printed in the business center. Spread out on the kitchen table and counter, System had several screen captures of Google Earth shots of 11738 Robinson Canyon Road.

Joe looked them over. "Seriously dude, when this is over you should marry that girl. She's pretty smart."

Naqeeb grinned. "No shit! Ya think?"

Now that they had a bone to chew the two got into it. The pictures showed the house set far back from the road. At least half a mile. The pictures helped them lock down where the road leading from Robinson Canyon to the house. The simple cattle gate was totally unassuming. The pair had driven by it many times the other day, unsuspecting.

Other shots showed the house and garage plainly. The house was a typical Carmel mountain vacation house. Wood logs, stone, fireplaces, windows. A "U configuration allowed for a circular driveway and a fountain in front. Deck and pool off the back. The roof was slate, however. Fire safety. Several stumps showed where large Douglas firs and pine trees had been cut down. It's all about defensible space in California.

That also prevents anyone from sneaking up on the house, Joe thought.

The garage/workshop was newer than the house. The two-story rectangle with a large roll up door on one end and cream colored aluminum panels to complement the green roof. There were clear sight lines for 300 yards to the back forest and the 30 or so to the house.

"Look here". Joe pointed. The high angle shot showed the front and the two sides clearly. The right side away from the main house had two large concrete pads holding utility units. Air conditioning, generator, and ventilation units. Large tanks were also visible going back along the 150 feet length of that side. The left had small pads indicating smaller doors into the shop. Various vent pipes emerged from the top. Plus the solar panels. Lots of solar.

"That looks like it's completely off the grid, Joe said. Look you can see the house utility hook up here. But not one for the workshop. "I bet one of those tanks is water too, he said.

Naqeeb agreed. "Septic?" he asked.

"Probably, Joe said. The less the county is involved the better for them. No meter guy coming in, rural mail delivery in town. No one to bother them"

"What about this?" Naqeeb asked pointing. Joe studied the picture.

"Dirt mounds? He asked taking his best shot.

His friend nodded. "That's what I think they are." Three huge mounds of earth rose 20-30 feet and were 80 feet wide at the base of the hill. The mounds ran right at the tree line.

The men were silent for a minute thinking.

'Someone put in a basement", Joe said at last. "That's why the trucks were going in and out. They were hauling away the dirt."

"And bringing in equipment and finishing materials," Naqeeb added.

They just mounded up the last of it, Joe surmised.

Naqeeb scratched out some volume calculations. "We could be looking at 50,000 sq ft. underground."

"Find out if DCS ordered an elevator in that time frame." System spits out the purchase order in minutes.

The Thysson Krupp unit was delivered right to the house. "For use up to 5 stories", they read on the website.

They mulled over the ramifications.

"We gotta figure 4 stories right?" "Say, 12,000 per floor."

"Sounds about right to me," acknowledged Naqeeb.

"You know this kind of facility, Joe told him. "Think about the lab at Stanford and lay it out underground."

Taking a pad and paper Naqeeb wandered around the room muttering Yeah, Yeah, and sketching.

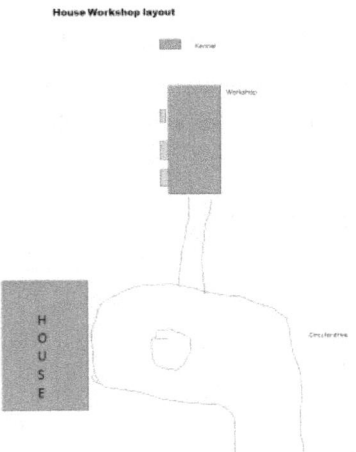

House Workshop layout

Carmel complex 3rd level

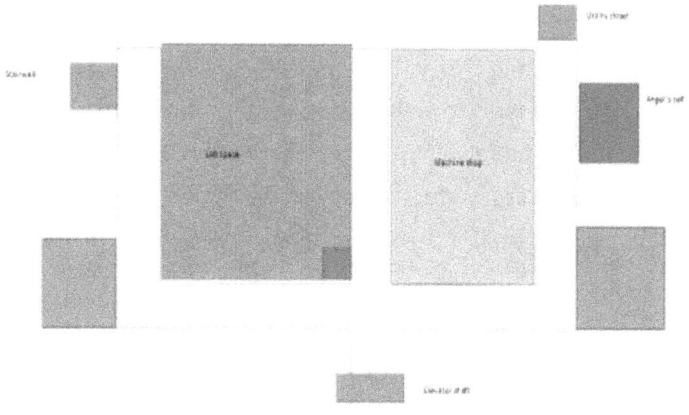

TOP View

Watching him sketch Joe became more depressed. *That looks right*, he thought. What's more, it felt right. He even agreed to where Naqeeb put Angela. A closet sized room as far away from everyone as possible. As far away from freedom as possible. Up and down one maybe two sets of stairs a day as they moved her from lab to bedroom. Shit!

Two escape tunnels? That bit felt fanciful but who was to say.

But the major thrust of the sketch showed the problem starkly: If FastFac truly did have 20 security and 20 scientists like System predicted, in the place then getting in and out was going to be impossible. The elevator was surely badged and guarded.

Hell, the Fairmont required a key card to get up beyond floor 7. Two guys could hold off the elevator from anything under a full scale military style raid.

Naqeeb finished drawing and looked at Joe. "This looks bad." He gestured. Plus this could be way off. "It's all guesses and conjectures on our part."

Smithson nodded. "Yeah, it looks daunting at first blush, but we need way more data before we throw in the towel."

His friend felt a surge of emotion due to Joe's loyalty.

They started discussing options and what data they needed.

"We need to know exactly how many security guys and who they are as well as geeks and who they are. We need schedules and layout of the buildings on a much more detailed level." Joe ticked off the items. "We need confirmation of where they are keeping her. Which level?" We need to know the security measures they have in place and figure a way around them." He sighed. "Then we have to come up with a plan and get her out then bring down FastFac and Margaret Pullman."

Naqeeb closed his eyes and blew out a breath. "Yep".

The two got down to work.

The weekend saw the men printing dossiers on the personnel System felt were at the facility. 38 people. 20 scientists and 18 security types. They started hanging out in the take out places per Systems recommendations. They did glimpse the FastFac security people. Following them only led to the gate. Not beyond. They drove over to DCS but found it locked down on a Sunday evening. An 8-foot chain link fence topped with wire was keeping all but the birds out. They went back to the gate and parked 300 yards beyond at a bend in the road.

Joe kept vigil in the car while Naqeeb clambered up the slope into the trees. He returned about 10 min later. "No luck" The same chain link fence is about 60 feet or so into the trees. It runs as far as I can see."
Joe drove off back to the rooms. "You think they have the whole 210 acres fenced?

"Yeah."

"Did you see cameras or motion sensors?"

"A camera," Naqeeb said.

"System is not seeing any video posted from that TV camera posted on FastFac website or digital storage locations, Joe mused. Closed circuit? Internal only?"

"That makes sense", Naqeeb agreed.

"210 acres is big, Joe said driving fast. I wonder about that backside."

The general area here off the canyon abutted the Ventana wilderness. There were houses scattered and the remains of old logging activities, but mostly the sprawl had been contained. The men were driving a 78 mile "C" to get to the front of the house. Their hotel room was only 14 miles from the back of the property as the crow flies.
Joe eyed Naqeeb. "Feel like a hike Tuesday?" He did.

All Monday was prep day. Buying supplies and planning. Naqeeb used System to book another week at the resort. He and Mr. McClain had further business in the area. Joe hit REI in Carmel for boots, packs, binoculars, maps, compass and more.

Naqeeb returned from the front desk to get busy using System and Google maps to figure out the best route. The wilderness area itself was vast but the protected lands ended just north of them. A mix of logging roads, private holdings, and public land showed on the maps. He finally zeroed in on White Rock Ridge road. That road got them close. Close as in 6 miles. Calling it a road was generous. Getting from the end of the road to the back side of the FastFac property was going to be tricky.

Joe came in while Naqeeb was printing out the maps. He put the gear on the couch and told Naqeeb to try on the boots. "They're twelves."

"Excellent, he said happily to be engaged in something, anything to get Angela back.

Putting on the wool socks and boots, he flexed his toes. Yeah, these are good. Joe had laid out the rest of the gear. "Any luck on the route?"

Naqeeb grunted. "Some." He walked around the room trying to break in the boots. Joe bent over the photos.

Naqeeb pointed. "This looks like the best way. We just take the 1 north towards Garrapara State Park. Just south of the park we hit Palo Colorado road. Right on Palo Colorado. There are some houses in the area but its mostly old logging roads." Palo Colorado connects here with White Rock Ridge Road." He paused. "I have no idea of the signs or the road conditions or…"

Joe held up a hand. "Not your responsibility, dude."

Naqeeb continued. "Anyway, we take the road to right here. The end point. That puts us 6 miles from the target."

Joe frowned. "That's the tricky part hey." A) We got to get to the place, and B) we got to get back in one piece."

12 miles round trip. The pair was in decent shape but this was no stroll in the park.

Naqeeb acknowledged the difficulty. He laid out the pictures and showed the preferred route.

"From the end of the road, it's a mile and a half until we intersect this stream: Salsipuedes creek. It is uphill for about a mile of that', he grimaced saying it.

Joe shrugged.

"Once we get to the creek we follow it to this bend point here," he pointed. Then it's due east by the compass for 4 miles to the back of the property." Joe looked on. "What if we miss?" Well if we hit Christmas trees, we are too far north, if we hit Robinson canyon road we are too far south, Naqeeb told him.

"Christmas trees?"

"Yeah, there's a tree farm north of the FastFac place."

Joe grinned. "Okay. We have pedometers, maps and compasses and good gear. We get in, scout out the area and get out, right?"

"Fuck em," Naqeeb said.

CHAPTER 19

June 23 0430 am 2010, Tuesday Ventana Ca.

The beeping woke him. Naqeeb rose and clicked off the alarm. He'd slept fitfully last night. Despite Joe's "Get in Get out" speech yesterday, Naqeeb had insisted on a trial run near the hotel. The trial was simple: ½ mile, turn left, another ½ mile and repeat and repeat. That should bring them back to point A. The reality was more difficult. They missed the trail head by 35 yards. They just couldn't see it through the trees. Only stumbling upon a group of hikers had allowed them to get back. Joe summed up the lessons learned: more frequent compass checks. The google maps were fine for showing detail but not elevation gains. The geo maps did that.

"And we need to use more natural landmarks to guide us" he finished.

Naqeeb dressed quickly. Joe was up. They slung packs into the car at 4:45 am. The day's first difficulty was apparent. Fog. The marine layer was in with a vengeance. The car's headlights barely penetrated the gloom on US 1. Joe missed the turn off for Palo Colorado. There were no street lights and not many signs. They hadn't passed a car since turning off the 1.

They found White Rock Ridge road in all its bone rattling unpaved glory. The sun was up but the fog was not giving up easily as they exited the car at 6:32. Flashlights on and packs ready they set off into the forest. The fog dampened sound and wetted the entire area. A thick carpet of needles cushioned the forest floor as they ignored the "no trespassing" signs and headed in. The needles also hid roots and dips in the ground. The route was steeply slanted up as Joe sighted with the compass.

"See the trees that make the "V"?" I make that North."

Naqeeb agreed and they set off. 10 min in and they could barely make out the car. They still had a position on the google map smartphone app. Another bearing and another landmark and they continued in struggling up. 40 min later they took a break. 700 yards in. Both were winded and wet.

"We knew this, Naqeeb reminded him. Salsipuedes creek is formed off the watershed on the other side of this mountain. I make it another 1500 yards to the crest."

"Plus it levels off some before we hit the true crest," Joe said trying to be positive.

The pair moved on seeing no one. Another 90 min of hiking and they hit the top. Taking a break to drink and eat a power bar, They checked GPS. The signal was gone.

Sighting an outcrop of rocks that was north the pair made their way along the side of the slope. It began a gradual down gradient. They made better time and came upon the marker rocks. The sound of water was music to four ears. Down to the creek and splashing across it, spirits high as the first leg was accomplished. The creek nestled in the valley of two slopes. The forest was thicker here near the water source. A half mile of sidehill traversing towards the north brought them to the defined bend.

A wan smile from Joe. 4 miles and almost three hours to get to this point. And the tough part was still ahead. No more GPS signal and google maps.

On the positive side, the fog had lifted and the day was warming up. Both men drank copious water and pissed. "Secure your stuff. Quiet as possible from here on in."

Joe went over the reminders. He looked at the slope in front of him. Again heavily treed and steeply angled the other watershed for the creek looked daunting. "We have miles to go.

"At least two miles to the top of this slope, Naqeeb told him. Then another two to get to the property. It's relatively flat up on top, though."

Joe nodded. "Come on man let's get to it."

The men started up taking their time and constant bearings to landmarks. This side of the valley caught more fog and rain so was a better home to bushes and vines versus a straight tree filled slope. It was tough slogging. 2 ½ hours of solid effort got them to the top. Scratched, blistered, tired and winded.

"Man, I used to love blackberries Joe said, but those vines are killers!"

Naqeeb agreed. "Plus I'm worried about bears, he said.

"Too early, Joe opined. Not much fruit yet.

"Plenty of thorns though."

"Fuck yeah."

"Should we climb a tree? Joe asked, "See what we can see"?

Naqeeb didn't think so. "We're are still at least a mile and a half to the property."

"Arghh, Just go!"

The small ridge top plateau offered the most open hiking they had seen. The widely scattered trees allowed for bearings and easy hiking. They covered a mile in 20 min.

"Man, we needed that", Naqeeb said huffing.

Joe climbed a tree about 20 feet high. To the east, he could see neat lines of Douglas fir and pine trees set out. Excellent! The Christmas tree farm! Close. "We go into the Xmas trees a little ways then walk due south until we hit the fence line, Naqeeb said.

"Sounds like a plan."

"Remember to look for motion detectors or cameras on the fence line when we get there."

Joe didn't agree with motion sensors. "Too many critters," he thought out loud.

The men went on. It was afternoon. They went into the neat rows of trees about 500 yards when Naqeeb turned them south. Another 1000 yards brought them to the FastFac fence.

The same fence as the front greeted them mockingly. Joe and Naqeeb stayed well back in the light screen of trees as they marked the fence. There were indeed cameras but they were inside the security fence. Mounted high on trees every 300 yards or so the cameras pointed along the fence line.

"No motion on the cameras, Naqeeb said observing. "They just use them to monitor the fence area and keep out everything."

The lack of sweep and the angle of the cameras gave them latitude to approach pretty close. It was 2:00 pm. They needed to start back if they wanted any chance of getting to the car before full dark. Lucky for them the days were so long right now. Daylight lasts until 8:15 or so. The last thing learned at the fence line where the guard patrols. A man swept through leading a dog on a short leash as he walked the perimeter. Shit!

Demoralized the men started back. At 8:35 two figures stumbled out of the trees into what the GPS said was White Rock Ridge road. The gloom of night and the fog rapidly turning the area dark dark dark. The pair walked up the road and around the bend. The car sat placidly, untouched. Loading in, Joe drove home in silence.

Naqeeb came awake slowly Wednesday morning to the sound of the front door closing. Joe had breakfast sent in. His friend was finishing his food when Naqeeb rolled out of the shower.

"Sore?" Joe asked as Naqeeb sat and started spooning food onto a plate.

"Bad," he said sounding down.

Sitting at the table he ate mechanically. "We are fucked," he said around the eggs.

"We can't sneak in and we can't shoot our way in, he told Joe.

"I agree," Joe said at once. "We need help." The statement hung out there.

Naqeeb chewed. "Who?"

"NSA," Joe said quietly.

"They take us straight to jail!" Naqeeb said at once.

"If that's what it takes to get Angela back, so be it." Joe intoned.

Naqeeb turned to his friend tears in his eyes, "Thanks, man!"

"Hey, jail is plan B though right?" I got an idea on plan A." He laid it out.

Naqeeb paused eating. "Okay, let's work it."

7:51. Joe booted up, ready for System. They needed to catch a break. Neither man looked at the Giants score.

CHAPTER 20

June 26 0810 am 2010 Friday, Rockville, Md.

Sipping the coffee at her desk, Stacy Brown read the sports page. Nationals lost to the Giants to allow the sweep. Damn the Giants looked legit. The phone rang and she answered immediately. "Yes." A series of clicks responded. Stacy waited it out. 'Code 22" a voice said. "Connect" Stacy responded. Well, well.

Click. "Yes," Stacy said again.

"Stacy? Joe Smithson." The voice said.

"Joe I'm surprised and pleased to hear from you," she said in response.

"Yeah, I'll bet," Joe answered.

"You dropped out of sight, Stacy complained. "I'll bet your mom misses you." More clicking on the line.

"Stop, Joe commanded. "I'm in trouble. Big trouble."

"What kind, Stacy asked alerted.

"I need your help," Joe told her.

Stacy hesitated. This was not standard protocol. "I'm not sure we can…" she began.

"Here's the deal, Joe interrupted. "I'm sending you a file. Call it a gift and my bonafides." You can do what you want with it." "I need help and I have a bigger gift in return for that help," Joe concluded swiftly.
Stacy processed this in silence.

He took that for assent. "Good. I need you and the backup band out here by tomorrow night. "I'm sure you have the location of this landline by now. 5:00 pm prompt please, we have dinner reservations. And don't bother calling me a pain in the ass. Bye."

Stacy waited until the clicks stopped. "Standard landline location trace, she said into the line. "Affirmative" was the only response.

She leaned back thinking hard. Smithson? Trouble with him she could believe. Had he been approached?

She turned to her computer. Logging into the isolated external e-mail account, she found nothing. Her internal account, however, had a new message. Fuck! She started isolation protocols. She detached the laptop from the network and ran every bit of virus and security software the NSA owned. Nothing. Something simply waltzed in and dropped in this file. Was Smithson's program that good? Trouble?

She transferred the file to a dead laptop and set aside her other machine. She isolated the e-mail file and opened the message. She read through twice. The attached file was pages of data. She was only a few pages in when the phone rang. Her head was spinning as she answered.

"Yes". More clicks.

"Trace complete. Payphone located outside the Heartbeat Gallery, Big Sur California."

"Thank you." What?

She clicked through to another line. "I need a meeting with section Green head ASAP. Affirmative.

Stacy had work to do. She spent an hour corroborating the Smithson file. She didn't want to dig deeper. It may trigger something and tip off the agent. Just before her meeting, she called Knowles.

"Mary? Stacy. I need op research. Hotels within a 50-mile radius of the Heartbeat gallery in Big Sur California. Guest registration. Three rooms maybe two or even one. White male, Indian-American male, and Asian American female all early twenties. She gave the names. The ABC and the Indian may be together. Find them, please. Thanks"

Her meeting went well. She laid out her case and the green section head exploded.

"You've had a hard-on over him for years," the man groused.

"You told me to come back when I had solid evidence. Now I have that in spades", she said a touch too smugly.

"And where did you get all this?" Green Head asked.
That question also opened her meeting with her own section head. Blackhead pinned her with it.

"A new source has opened up. Civilian computer whiz in California. He promises more. I need to take a team to California and lock him down," she admitted. She said nothing about helping Smithson.

"Go get him," her boss directed.

Stacy wasted no time contacting her team. She wanted to be out of the office by 8:00 pm so she could get a good night's sleep before the op. The company plane was leaving early next morning. She spent three hours drafting requisitions for people and equipment. She drafted her team carefully. Thomson, Rhoda, and Kusick. She'd worked with all of them before. Smart professional family men. Steady. She sent instructions: pack for a week. Draw equipment at 0600, wheels up westbound at 0700. Her last line: Get a good night's sleep. Brief on the plane.

Mary called back as she typed. "Stacy, it's me. Weird stuff going on here. There are 25 hotels in the target area that match the criteria you laid out."

"That's impossible, Stacy rejoined.

"Tell me about it. 12 of the rooms show various levels of charges, Room service, and spa stuff. Five more have requests for additional items, roll away cots, extra pillows. Two of the reservations end today while six more starts tomorrow. Some of the names are repeated but they are mostly new. Chinese woman, an Indian man and a WASP man." Mary sounded exasperated.

"It's like the person knew exactly what kind of data screen we would be running and is providing a real time smoke stream of results." Mary continued. "Of course that's impossible.

"So the only way to verify any of this is to physically send someone to each hotel to get eyes on the occupants, Stacy said. She knew the expense and trouble that would be to her. "Pain in the ass" she grumbled under her breath.

"What?" Mary said.

"Sorry, Mare. Him, not you." "Please print off all the material you have and I will review it tonight or on the plane," she asked. "Thanks, Mary!" She hung up.

"Smithson, what the fuck are you doing?"

"Dude, I hope you know what the fuck you are doing!" Naqeeb's voice sounded perturbed over the drop phone.

"I do. Just proceed like we planned," Joe told him as he got out of the car. "I'll text you from the road."

He emerged from the trees about 20 yards from the pay phone outside the gallery. A few tourists were hanging but the place was pretty quiet for a Saturday afternoon. And it was a beautiful day to boot. Joe walked towards the woman with a smile on his face.

Stacy had a scowl on hers. "You could at least fake being happy to see me? Joe told her. Stacy put a finger to her ear unconsciously and the scowl deepened. Joe stopped leaving a few yards between them. He didn't want her to be able to break his neck. Arms spread wide he said, "as your boy must have just told you Naqeeb just bought him a beer at the Ventana and the two are happily chatting about the Giants."

"How in the fuck did you figure that?" Stacy asked.

"Did you like my present?" Joe asked in turn. "You must have, you're here aren't you?"

"How in the fuck did you figure that? She repeated.

"Look, we have reservations. Let's get back to the hotel. I'd rather do this once for your whole team instead of three or four times." He looked at her. "You want to signal in the other two guys? Let's take my car back, okay?" Stacy scowled but texted. "I promise to explain once we are all together. Then it's up to you." The pair walked back to Joe's new/old car. The first 5 min were in stony silence.

"It was the bitcoin that gave you away", Stacy finally said.

"Yeah, my predictive matrix said it was an outlier. Only the Post Ranch Inn and Ventana accepted the bitcoin." Joe told the half-truth with a straight face.

"I have to tell you that 4 hotel managers swore you were guests even though they had never set eyes on you," Stacy ground out. "Did your program set all that up?

"Yeah, we've been using my program a lot lately", Joe kept it as general as he could.

"Why?" She asked him bluntly.

Joe was silent.

"You asked for this meeting", Stacy reminded him. "Well, here we are!"

"Yes, I did… Joe started. "But suddenly…"

"Suddenly you're not so fucking sure?" she said coldly.

'Suddenly I just realized why my sister calls me the dumbest smart guy she knows," Joe admitted shakily.

"Look, I told you I needed your help, he tried to start up again. "You caught the embezzler, Owens, right? Joe looked at her as he drove up the twisty road.

"You tried two years ago to get him but you didn't have any hard evidence, so it didn't stick right?" Stacy said nothing. "Well, we used my program and got you the evidence." He finished.

"How? Demanded Stacy. "Your program did not have the ability to crack encrypted files last I checked." "How did you penetrate Owens files to get that detailed bank records?" Not to mention getting the file to me."

"There have been some modifications to my program," Joe lied badly. "Bottom line we got you the evidence you needed," he started again. "I have a bigger fish to give you in exchange for your help." Joe finished. Silence.

Stacy looked out at the ocean for a time. "What kind of help? She finally asked. "I need you and your team for an op." Joe regretted the words as soon as they were out of his mouth.

"An op?" Stacy exploded. This isn't the fucking A-Team you fucking idiot!" "Start telling the truth!"

Joe looked abashed while driving thru the redwoods near Julia Pfieffer Burns State Park. "Naqeeb and I can't do this by ourselves. We need your help. And you have the skills. I mean Lynette Childers won the bronze Star in Afghanistan in 03 attached to Army Intel." Joe looked at her. My god, what were you 19?" "Wounded twice. 2 tours and then poof you disappear. I mean, Lynette disappears."

"Never call me Lynette again, ever, Stacy said quietly. How? She processed how it was possible Joe had this information.

"I told you my program has gotten an upgrade, he said. Two cars trailed him as Joe turned into the parking lot at Ventana. Joe parked. He turned to face Stacy.

"They have Angela and we need you to help us get her back."

"Who has her?" Stacy asked.

"FastFac, Joe answered simply.

"The company?" She said confused again.

"Yeah, Margaret Pullman grabbed her to put pressure on us to hand over my program. System we have been calling it." Joe gripped the steering wheel.

"Move out", Stacy directed.

They got out of the car and marched into the restaurant through the hotel lobby with Stacy's men flanking them the whole way.

"Good evening Mr. Mc Clain", said the hostess.

"Good evening Bobbi", said Joe.

"Mr. Chopra and your other guest are waiting for you if you will follow me."

She led them to a front window table with an incredible view of the ocean and the setting sun. Naqeeb and the NSA man were sitting eying each other like cats. Joe winced. The four newcomers sat and waited silently while menus were passed around. The waiter dutifully went through the specials. Joe ordered two bottles of wine. A red and a white. More silence while glasses were filled.

"Teammates" Joe offered as a toast. No one drank.

"Lay it out," ordered Stacy.

The men did so. Joe started with a highly edited version of events starting with the break-in at the apartment.

"I guess FastFac got wind of my program, System and I decided to grab it." He said. "Only they got the wrong laptop." "Plus they roughed up Naqeeb and Angela in the process." Joe related.

"Why not call the cops? Stacy asked.

"And tell them what? Someone tried to steal my program that Stanford owns and I'm not supposed to have?" Joe said urgently.

"We lit out for Naqeeb's Aunt's place in the city to regroup, Joe said. One of the NSA guys shook his head at that revelation.

Salads arrived as Naqeeb took over the narration. "We knew we needed to be off the grid so we went to a hotel to try to get FastFac to back off. He went over the denial of service attacks and the anvil folder while skipping Kocinski's money.

"It took a few days but they agreed," Naqeeb said.

Joe noticed the NSA people were eating and drinking. That's a good sign, right?

"We thought we were home free he confided. I actually went back to the apartment. Naqeeb and Angela stayed at the Fairmont to celebrate and that's when they grabbed her. They tried to get Naqeeb as well but no joy." I guess they figured one was better than none."

One of the NSA guys, Joe thought his name was Thompson asked, "How did the take her?"

Naqeeb related the story with the note and his escape to the youth hostel.

"Punched her huh? "Rhoda," said. "Lucky".

"Smart also," said the third, Kusick. Stacy stayed silent. Dishes were cleared while they told of the search for FastFac facilities that led them here.

Joe pitched what they needed. "We are 99 percent certain that FastFac has her in this complex in Carmel Valley." Naqeeb pulled up a three ring binder that had some select data from System to show the NSA. 'We think the main lab is underneath the workshop." Joe passed around Naqeeb's crude drawing. "We think a local company called DCS Construction did the work." He showed the aerial views with the dirt mounds. "We need to get her out of there," he finished. "That's where you guys come in. We can't do it ourselves."

Now Stacy spoke up. "And why should we help you? In exchange for what?"

"Because one of your people is passing secrets to the Russians," Joe told them flatly.

The NSA people stiffened. "This is win-win, he told them. You help us and I give you the name."

"And if we take you to jail and torture the name out of you?" Stacy asked. Her team looked perfectly comfortable doing just that.

Naqeeb started to object but Joe grabbed his arm. "That would suck, he told her. But I ask that you get Angela out anyway, No matter what happens to us she needs to be rescued."

The silence stretched. Naqeeb strained under his hand. *Be cool* Joe prayed.

"I want to see more of what we are dealing with here, Stacy said reluctantly. "I make no promises not to haul you in after this is over."
At least two people around the table started breathing again.

The waiter slid in with the check and Joe signed for dinner charging it to the room. "Let's go back to our room and we can go over what we have and answer your questions." The NSA team agreed. Stacy made a discreet call on the way. The grim procession worked its way to the room with minimum chit chat.

Rhoda whistled when he saw the suite. "I like the way you guys operate."
The six arrayed around the kitchenette table and counter sipping 12 dollar mini bar beers.

"We start basic," Stacy commanded, "how long has she been gone"?

"Two weeks today," Naqeeb told her helpfully.

"Do you have the ransom note or e-mail," she asked.

Naqeeb was stumped. He looked at her not knowing what to say. Joe jumped in. "We've had a call from them and her," he said hastily. "They just said deliver up my program, the System and we would get her back." They haven't said anything about the exchange. They let us talk to her for a few seconds and then she's gone." He finished lamely.

"That makes no sense Stacy complained. "Two weeks with no demands? How are they keeping her absence from friends and family? Work?"

"She told us she has talked to her family and Stanford. She told them she was working on something and would be out of touch for a while. Naqeeb provided.

Stacy and the NSA guys exchanged looks.

"That's kind of how she works, Joe offered. Everyone looked at Stacy who frowned.

"How do you know she's here?"

Joe laid out the cell phone calls and the corroboration with the FastFac property via Marcus Baxter. He showed the team the DCS permits and the neighbor complaints.

The team poured over the records and the drawings along with the invoices for equipment. The elevator especially. Lastly, they went over the personnel dossiers of people they thought were there. The take out bills were taken note of.

"How many do you think are in there, Kusick asked sitting back?

"20 or so security and 20 workers, we think," Naqeeb said and Joe nodded agreement.

"No sneaking in there, Kusick said sitting back. Naqeeb noticed something. All 4 of the NSA people had had the exact same amount to drink: One-half glass of wine and three tiny sips of beer. "It gets worse". He detailed the fence and the cameras with the dog patrols.

"You hiked all the way in and back?" Thompson said.

"I love her" Naqeeb said simply.

All the NSA people shook their heads "no'. That's not how professionals acted. Stacy reviewed all the data. "And your program put all this together?"

"System is amazing," Joe agreed.

It boiled down to Stacy. "Okay. We are going to help you in exchange for the name." All business now. "What are our resources? How much cash do you have?" "About 6,000 Joe told her relieved. "We have a car and some hiking gear. We can access more cash if we have a couple of days to build up and we have about 80,000 on the bitcoin wallets."

She seized on that. "That's how you are affording this luxury?" Joe nodded.

"I'd like to see that trick work." Stacy directed.

Naqeeb ran her thru the matched pairs on the bitcoin exchange. Kusick obviously knew some things about forged credit cards and charge clearing houses. His questions were right to the point. Joe winced when Naqeeb popped the USB thumb drive into Systems laptop and showed his fake I.D's. Soon five heads were glued to the screen while Naqeeb made the NSA room reservations. System was blessedly quiet while this process was run. Naqeeb mostly used Expedia, not System. They had to use the Nano when rooms needed to be shuffled to arrange for groups and families to cancel their vacation plans. More than one plane had been grounded. Joe felt sorry for that. The Ventana was having a weird summer room wise.

Kusick cocked his head. "You might have some problems if the exchange does a zero balance audit on a quarterly basis. It's July 1st. They may have done one yesterday or today." "Bad for you" he concluded.

Joe gulped. System had alerted them to just that fact as he went to get Stacy. System was quiet because the Nano was steering the audit right around their holes.

"Thanks, man, I think we are good."

"How long have you been on the run total, Thompson asked.'

"Three weeks altogether, Naqeeb admitted.

"Feels like three years, Joe added. This time the head shakes were knowing.

Stacy shuffled the papers until she came to Naqeeb's crude drawing of the underground complex. "We need hard data, not guesses. Thompson, get two thousand from Joe. Rent a helo and take up the infrared gear. We need to know who is where."

"Four stories is too deep for the gear, he reminded her.

"I'm betting that the senior security people and the head scientist occupy the house while the majority of the staff and a few of the security guys live in the garage top floor. Kusick, Rhoda, you guys are on the DCS files. Find the architect from the job. Target his personal computer. Nobody deletes everything," she said authoritatively.

Joe winced. We should have thought of that he thought.

"What can we do?" Naqeeb asked eagerly.

"Sit right here, Stacy told him. "I'm keeping my eyes on both of you."

Shit.

CHAPTER 21

July 3, 1147, am 2010, Monday Mountain View, Ca.

Cantrell finally had time to look over the Sunday Chronicle business section article they'd planted. "Teaching an old dog new Tricks", was the title. That editor should be fired, he thought as he read the article. It detailed all of FastFac efforts over the last years on innovative new products.

The pure research of IBM or the GE Labs was old news. Applied research in targetable areas with tangible results was in. All in all not bad. The stock was up three percent today.

Cantrell checked the ticker. He breathed a sigh of relief. They'd had two weeks to right the ship. Ever since the Chin woman had been grabbed, no more denial of service attacks, no more website issues and messages were getting routed. Smooth.

The Carmel office reported remarkable progress. The Nano agents were linked and suspended in a fluid medium. Cooper was astounded. He regarded Chin as a cross between Madam Currie and Mother Theresa. Amazing what a little confidence got you. The market was beginning to react to their constant hinting of a new direction. The fact that Peg made a pretty public buy of shares was seen as a very key sign. Optimism was contagious. The stock stabilized and started the long climb out of the hole. The campaign was benefiting also. Pullman's poll numbers were up. Still 18 points down but better.

Peg was better as well. More stable, more rational. Less homicidal, if Cantrell allowed himself to be honest. As he noted the Giants victory over the arch-rival Dodgers on the sports page, he reflected that her drinking was more manageable. The manic cycle was a great high if you could ride it out.

The blurb on Kocinski was buried on page 14 of the local section. *Stanford Professor killed in a hit and run.* A loose thread was snipped. Cantrell could feel it. They were close now. All they needed was one picture of the Nano agents. Sentient or not the mere sight of the complex agent bodies would send them to victory!

He needed to communicate with Cooper. The scientist had to get them that picture! A clean pic was worth a thousand words. A thousand thousand thousand dollars more like! Cantrell was sure that as soon as Peg had that picture in her hand's more loose ends would be snipped. Yeah, he thought to pick up his pen to write to Cooper. I don't think Chin, Al-Meri or Smithson are long for this world after we get that picture.

"You promised me!" "I've held up my end of the deal!" Angela screamed through the door. The security people had locked her in here after her argument with the lab folks.

The day started normally, with her minders bringing her up to the lab and then to the conference room for a status meeting. Normal stuff as far as that went. Cooper wanted pictures of the Nano agents. Corporate was demanding them, really. Angela was up to here with Cooper and his willingness to turn a blind eye to her captivity and take all the credit for her work.

"It doesn't work that way!" she said. "A) they have only had a short time to link, the Nano's may not be in the full lattice cube bodies as we've come to know.' "And B) Your imaging system sucks."

The two assistants that worked most closely with her kept their heads down and mouths shut.

"Ms. Chin, Cooper began.

"Fuck you, doctor!" Angela snapped. "What are you going to do kill me? Steal my work?" "I've done my part. The Nano agents are growing. "If you aren't competent enough to get pictures that is your problem." She withered him with a look.

Cooper blushed and signaled the security goons to take her out.

Again she reacted atypically. "No! You owe me a phone call!" "I get to talk to my family and Naqeeb today!"

They dragged her out and down to her "suite". That was all of the facility she had seen during her time as a captive. The lab and her room. She had cooperated in order to make the nightmare stop and then out of boredom. Now she was pissed. "Assholes" she hurled at the closed door.

Twenty minutes later the door opened and her two personal escorts stood there. One extended the phone for her to use. 'Parents first," the man said. He hit the send button and the phone dialed. Angela had the phone which was already on speaker.

Her father answered.

'Hey dad, it's me, she said right away.

"Angela, he said, I'm glad you called. I got a call from Ed Lee's office. They are interested in that BART improvement project you sent."

Angela closed her eyes. No "How are you?" no "We've missed you", just straight to business. She suppressed tears. "That's great daddy but I'm super busy with work and I won't be free for a month or so," she said willing him to ask questions.

"No problem, he told her jovially. "It'll take years for the legislature to act anyway."

"How's mom and Suelyn?" she asked desperately.

"Fine, shopping," he told her. At least they were all safe. The security guy squeezed her arm.

"Well, I just wanted to check in and see if you are alright."

"We are all fine here!"

"Okay, I'll see you when I get the chance, she said softly. 'Love you"

"Love you too sweetheart," her father said and hung up.

Angela took a moment and composed herself. No help from home but at least they were safe she kept repeating to herself. The security team dialed the second number and handed her the phone. "No funny stuff," he warned.

The four people in room 322 in the Ventana Inn certainly reacted when the drop phone rang. Joe and Naqeeb sprang up like a well oiled machine. Joe furiously typing into Systems laptop and the other racing to get paper and pen and the phone. Joe finished the trace request to System while Naqeeb furiously waived the NSA team to silence.

Naqeeb put the phone on speaker and answered. "Angela?" he asked in a shaky voice.

"Honey!" she cried.

"Baby, stay calm," he told her. How are you?"

"Fine, but I'm going batty, she got out.

Again the phone was wrestled away from her to the sounds of a struggle. With Angela swearing in the background, a muffled male voice said, "Stick to the deal and you'll see her soon." Click.

He swore and looked at his friend.

"Same tower on Robinson Canyon road."

Kusick shot Stacy a significant look. She shook her head at him.

"Batty," Naqeeb said to the room at large. "She's telling us she is being held in a cave. Underground." He finished off the thought as Thompson and Rhoda came into the room.

Heads swiveled to look at Stacy.

"We need to talk," she said authoritatively.

The six people held the come to Jesus meeting that the roommates had feared was coming. They could not keep System from the NSA any longer.

Anything to get Angela back, Naqeeb thought.

The NSA people all regarded the pair like suspects. "Okay, he said. "Start with the questions you all must have." He looked at the NSA agents each in turn. No friendly faces. A quick glance at Joe showed he was also squirming.

"What is Angela really doing at FastFac's complex?" Stacy asked.

Boy, right to the hard part. "FastFac took her to replicate her Nanotech work she did for Stanford," Joe put in quickly.

"Total truth, she said over him. "We know what she was doing at Stanford and the results." Stacy had obviously been talking to Ft. Meade. She knew some things, but not all.

"She did it," Naqeeb said proudly.

A look told him to provide more. "She designed and constructed a fully aware sentient Nano agent system," he said by route now. "We, strangely enough, call the agents "System".

Disbelief warred with certain facts in the NSA agents heads. How did they do that?

"What do you mean? "Fully aware", Stacy asked.

"Fully aware. Alive. Sentient. A new life form," Joe told her straight-faced.

"Holy shit," came from Thompson. The tall man was impressed.

Now the questions came forth from every angle.
"Is that how you tracked the cell phone call that quickly?" Did you use them to create the digital wallets and the bitcoin transactions?" Is that how you found the DCS architect when we had so much trouble?" "Is that how you tracked Angela down here and figured out who had her?"

The questions piled up while Stacy stared at the pair. The room grew uncomfortable. Joe hitched in a breath. "Yes." Simple, honesty is what they needed now.

Naqeeb sat Systems laptop in front of Joe. The NSA agents dutifully crowded around behind him so they could look at the screen. Joe grimaced.

"Hello System, he typed.

"Good afternoon, Creator Joe," System replied. Creator Angela is still being held at the FastFac facility, System reported. "All relevant personnel remain in last locations, however, System Kocinski is no longer operational."

"What?" Stacy asked as Joe and Naqeeb gasped. Smithson asked for supporting evidence from System. The Nano's flashed the Chronicle article, with a

"Creator Joe should know by now we can be trusted on these items", message.

Rhoda looked that over. "How does it know who you are? Is it tapped into a camera?"

Heads swiveled. Joe raised his hands. "No." "First off, it's they, not it. The Nano agents share a kind of hive mind. What one knows, all know. They always refer to themselves in the plural. Second, they are not tapped into the camera because this laptop doesn't have one."

"We think they use word patterns, typing speeds and idiosyncrasies to ID who is typing, Naqeeb added.

The NSA team read thru the Chron article detailing Kocinski's death by hit and run.

Kusick used his smartphone to verify the article.

"Who's Kocinski and how does he fit in?' Stacy asked.

Naqeeb led them thru the story in bits. Disruptive technology, going to him for help and the betrayal.

"So he ties FastFac to you?" Thompson stated more than asked.

Yep.

"Pullman and FastFac are cutting off leads that tie back", he mused.

Stacy thought that through a little.

"Do you mind", she gestured at the laptop where Joe was stationed.

"Uh…"

"Let me rephrase. Move"

He got up stifling the objection from Naqeeb.

"System?" She asked again getting the nod from Joe.

"System, pull all relevant data on known Pullman associates, friends, business partners, and competitors who have died suspiciously in the last 20 years." Stacy typed away.

"System will not comply," came the immediate response. Where is creator Joe or Naqeeb?"

Joe wanted to hug those little Nanobots!

"What's this? She demanded.

"What portion of self-aware sentient life form don't you understand?" Naqeeb said with relish.

"They are a little finicky about who they trust apparently, Joe put in.

The tension level rose in the room.

"Does it really matter who does the typing? Joe said. "We need to focus on getting Angela back before she gives FastFac and Pullman a System of their own. Imagine what a ruthless character like Pullman could do with System?" he said.

Thoughts turned inwards. The NSA team lead had seen how much information System had gotten, and had used it. She didn't think Washington would like a megalomaniac CEO to have unlimited information at her fingertips. If knowledge was power, System pulled a lot of weight.

"What happens after she is returned, happens," Naqeeb said looking at each person in turn.

Nods around as Joe slid back into the driver's seat. He repeated Stacy's basic query on Pullman's associates and strange deaths.

System provided them with four additional cases besides Kocinski over 20 years. Two men from the nineties. Both FastFac directors. One suicide and one car crash. The more recent ones were a Swiss banker no one had heard of and a woman. Both hit and runs.

Naqeeb stiffened as the picture of the woman came up. "I've seen her before." "She was part of the team that kidnapped Angela." He detailed the note and slugging Artimus. "This is not the woman I hit, this is the one who tried to taze me and the guy got in her way," he concluded.

"But she was definitely part of the team that kidnapped Angela, Thompson asked.

"Yes."

"So they killed her and now Kocinski", Stacy mused.

"He got him, Joe said fiercely. He sold us out to FastFac and tried to profit off of that." Good riddance."

Kusick had the two men go through the signatory documents and how System handled the brokers. At the end of the story, the NSA man was impressed.

"Slick. Yeah, System in the hands of a person without a conscience would be very bad," he confided to Stacy.

Naqeeb tried to refocus the team on retrieving Angela. "This isn't getting Angela back, he told them.

Stacy agreed. She looked at her team. "Thompson, what did you get in the way of hard data?"

Thompson opened his phone. The other 5 crowded behind him to see the infrared video he had shot from the helicopter.

The view started high over the main house.

"The camera confirmed 12 bodies in the house with 7 more visible in the workshop plus 4 outside. It's possible based on the position that 2 of those outside were workers and 2 security. "

Stacy paused as the camera view swung over the house to the workshop. "Did they notice the helo?"

Thompson shook his head. "Not more than a passing glance. We only made two oblique passes."

Indeed the camera angle rapidly swung to the workshop and then back to the house. There was perhaps 25 seconds of footage. The second set came from a back view and swung over the house and workshop back towards the road.

Thomson reset the video and slowed it down so more detail could emerge. The video showed as a shaky, grainy greenish image with bright white heat blooms in certain areas of the house.

"The direct overhead shot clearly showed areas of heat concentration" the agent, explained for Joe and Naqeeb's benefit. "I think these areas here are bedrooms. "They may be bunk beds which allow four to a room. You can see it's only 2 by 2 in the beds now." The roommates couldn't see that at all. The NSA man ran the video back and forth slowly pointing out features of the house while NSA agents nodded and Stanford computer geeks scowled.

"The workshop is harder to discern with the metal roof."
Oh, great thought Joe.

Again the video was played back and forth from both passes.

"You can see the cluster of four right here" Thompson pointed. "The other three are scattered."

"The outside people are teams of two with dogs." Thompson paused to make sure everyone was keeping up. "The kennel has two more for four total."

"Believe me, you'll want to know where those dogs are if we have to go in", Rhoda told them, in response to a look on Joe's face.

Kusick chuckled, "Remember, Prague?" The other NSA people shuddered.

WTF? Thought Naqeeb.

Joe bit back his reply. He really couldn't make heads or tails out of this video. Stacy nodded as Thompson finished up.

She looked at Rhoda. "Go".

"Just like you thought boss, he said. "House plans and workshop right on his computer. Full specs plus utilities and elevations", Rhoda finished with an evil grin. "I duped em without a trace." He held up a thumb drive.

Rhoda went to insert the thumb drive into Systems laptop. He wasn't thinking. Stacy was, however. "Hold!"

Rhoda paused, looking at her and realizing what he'd almost done.

Stacy turned to Joe and Naqeeb. "Will they infect the thumb drive if he inserts it?"

Shaking his head, Joe said, "No, they have to ask us before they can move anywhere."

Skeptical looks.

"Suit yourself, he said, but we've looked and they have kept every agreement we have ever made with them."

"Honest to a fault", put in Naqeeb.

Stacy gave Rhoda the go ahead. He inserted the thumb drive and Joe called up the plans.

The drive yielded beautiful blueprints for the house and workshop lab complex. Again the lay people had some trouble visualizing the layout. The elevations helped with that a little. Plus they did reveal that the lab was only 3 levels deep not 4 as the men had speculated. The size was 17,000 sq. ft. per level so the overall size was close to what they'd envisioned.

As the NSA people studied the prints with nods, Naqeeb glanced at Stacy. She was frowning.

"Can we try something? He asked her.

She used her most annoying/effective personal trait. She said nothing and stared at him.

"I think System could use the blueprints and superimposed the infrared footage on a 3 D model and help us better visualize what's inside both buildings," Again nothing from the woman.

She glanced at Rhoda and Kusick.

The latter shrugged. "Three to four weeks at Ft Meade", he told her.

Joe immediately started typing. "System estimated the number of cycles to superimpose the infrared footage onto a 3D model of the two building based on available architectural blueprints."

System responded that they needed to know the 3D programming and the files. 5 hours was the estimate after a few minutes internal processing.

Stacy gave the go ahead. While her team connected Thompson's phone to System, she slipped out to make a phone call.

Joe watched her go with apprehension. He knew she was keeping The Man informed as to what was going on here. He couldn't focus on that now. Angela first. NSA second. Ruin FastFac third. The rest of his life fourth. Whew.

Three rare hours of free time fell to them in the middle of the day. The NSA people worked out. Naqeeb and Joe tagged along. As soon as Stacy popped out 25 chin ups, the friends knew they were in trouble. They dragged themselves back to the room an hour later. Defeated.

System was true to their word... At 6:35 pm as dinner from the local burger place made its greasy appearance, the screen showed: System Ready.

"Here we go", Joe announced. The agents again crowded around the laptop. "Display results," he commanded.

System brought up a beautiful 3D model for the house, grounds, and workshop. The detail was outstanding. System used standard artwork for people and furniture and machines.

Stacy had Joe pause. "These are Systems guesses?" she asked.

Joe shook his head. "No. System will delve into an invoice to find out say the make and model of dryer. Then they will interface with the manufacturer to get a realistic representation of the dryer and show that in the model. So if System says there is a 60" Toshiba television in a room, you can be certain that is true."

Again doubts on faces. The camera view went from a high overhead shot to head height and looking at the front door. System now "Flew" the camera into the door. Grainy green tinged infrared video showed up here. System had not put any people into the entry way. Blue outlines peaked under walls to show where the blueprints were demanding that walls existed.

The layout became apparent even to Joe and Naqeeb.

The front entrance hall was relatively small with a wall and closet straight ahead. The 15 by 15 room was slate tiled and had wood paneling on the walls. To the left was hallway showing a series of rooms that System ID'd as an office, a utility area, and a bedroom bathroom combo suite. To the right was a large library and study with a nice desk.

Stacy ordered Joe to put up the raw infrared footage in an insert screen to allow the NSA to check System. Another insert brought up the blueprints. System obliged without a trace of emotion. The two creators were indignant on the inside to have their baby questioned. Still. Outside showed nothing to prevent pissing off the NSA.

The camera now walked the point of view around the short wall ahead of the entrance door and into the great room. This was the wow factor any real estate agent would want. A 60 by 60 great room with 20-foot ceilings dominated by a huge fireplace on the right. The back wall was the required floor to ceiling windows to let in the gorgeous views of the back patio and deck pool and the trees beyond that. The left side of the great room melded into the dining room with the kitchen beyond that. A small bathroom was off the kitchen and a huge pantry opened back into the hallway that ran off the left side.

Stacy had Joe pause the video again for more discussions. System had put two people in the kitchen based on the infrared. The ghost images of a man and woman showed. She quirked an eye at Naqeeb.

"Body heat differences, he guessed. Maybe height girth differences as well." Conjecture. Again, she was reminded: second guess System at your peril.

Stacy had other dangers in mind. "Remember this: If you go into the great room someone could run into the kitchen and thru the pantry out into the hall to sneak up behind you."

The men nodded solemnly, even Joe and Naqeeb.

She had Joe resume the video. A laundry area with a sneaky back stair showed up. The NSA agents groaned. None of them had picked up on that during the first go thru with the raw infrared and the blueprints. Surprise places for bad guys to come at you got people killed. *That's why we are doing this,* Joe thought.

The camera now reversed course and traversed the great room back to the right and into the mirrored hallway. Symmetry thy name is a beauty. This hallway didn't have a stairway but the rooms branching off left and right were mundane: Bathroom, gym, bedroom being used as storage area. System put one more person in the gym. Naqeeb shuddered as the camera viewpoint went by the male ghost.

The back of the entrance wall held two more staircases to access the upper floor. The hallway looked down over the great room and kitchen. "That's why this house is not 10,000 square feet, three-quarters of the upper floor space is wasted on air." Naqeeb meant to just think that but it slipped out.

Rhoda chuckled, "Yeah but it's the open, airiness that you want in your dream home!"

The upstairs was predictable, a big master suite on the left with its secret staircase down to the kitchen and more bedrooms and bathrooms on the right. System put 8 people in two bedrooms on the top floor. 4 per in 2 of the bedrooms. Two were empty.

Thompson couldn't contain himself. "That's the night shift security detail, getting rack time."

Stacy agreed but told him to hold his comments.

The camera now flew out the upstairs window and showed the grounds clearly.

The solar panels, the workshop, and the utility pads all evident. Heat blooms matched perfectly with the 3D representation as the camera moved to the workshop. The viewpoint flew into the large roll up door at the east end of the building. The infrared only went with this top floor. The roll up door led into a standard machine shop tool area. System matched up invoice items to layout and the blueprints. A small office area was near the door. The middle third held the complex utility machines and equipment a laboratory would need. The back third held bedrooms and bathrooms and the elevator entrance.

System and the infrared put one person in the office area. 6 people 2 by 2 by 2 in the bedrooms and another person standing next to the elevator entrance.

He was so obvious that as the camera passed by him Kusick murmured "Guard' under his breath and stiffened.

The camera continued into the elevator shaft and down. No more infrared. Just System and the blueprints. The first level underground was dominated by a large conference room and offices. The next level had two large lab complexes with a joining conference room. Each lab had a small lockout clean out entrances.
The lockout entrances confused the NSA men. "What? What is that for? Thomson asked." Naqeeb said, "It might not be full bio level safety 4 or 5 but I'm guessing at least level three clean room."

Stacy looked up sharply.

"How do you know that?" Thompson scoffed.

"Dude, maybe you should listen to the one person in the room who has worked at a Nano fabrication facility."

Thompson drew breath to argue when Stacy overrode him.

"What do you mean not full bio-safe level 4 or 5?"

Naqeeb suddenly went silent.

"What aren't you telling me?" the woman said coldly.

"The Nano's contain radioactive components," Naqeeb admitted.

Pin drop silence as the other four-eyed the laptop.

Joe rushed to fill the gap. "Yes the Nano agent bodies are composed of Technetium, Rubidium, and Radium contained within a carbon-silica lattice structure," he rattled off. Angela would have been proud. "However the amount of radioactive material is negligible." Joe finished up.

Several voices started to object when Naqeeb got up and started rummaging through the cardboard box that had seemingly followed them everywhere since the whole thing started.

"Here, he said flipping the Gieger counter at Thompson. "Check yourself."

Her men busied themselves checking out System while Stacy stared at Joe.

"I can't tell you everything I know about them in 5 hours." he retorted to the look.

He casually looked at his dad's watch. "In fact System also seems to need sleep. So in 35 minutes or so, they are going to shut down for the night."

"What!?!"

"Don't ask me, ask Angela and Naqeeb. Something about mechanical systems mimicking the natural world. So they put in a sun," he said. Questions built up for Stacy.

Kusick finally looked up from his smartphone and the counter.

"Angela said it was like 2 extra X rays a year," Naqeeb said softly.

"A little less, the agent confirmed. " I don't think System would set off the NBC detectors," he told his boss.

Stacy took the news calmly. "Alright, let's finish this off."

They clustered again at the laptop. The view now dropped to the lowest level. A large computer server room was displayed. Full component level electronics repair shop and utility equipment dominated this floor. The last section showed a small storeroom with a bathroom shower for industrial accidents.

"That's where they have her Naqeeb said squinting at the screen. He burned the location into his brain. No one disagreed.

The camera stopped at the end of the corridor. "That's a cruel joke," Rhoda said.

"What is?" asked Joe.

"Can you have System show the utility tunnel? He asked Joe.

"Sure". Joe typed.

The camera picked up again and flew thru the door at the end of the corridor. A small storage closet that had a second door opposite. The view floated thru this door into a small 5 by 5 opening. Various pipes and ducts ran into this room from both walls and the ceiling. The pipes all merged here and then ran upwards as the tunnel rose out of the opening. The tunnel had brackets every 10 feet or so to hold the pipes ridged. The slope didn't appear too bad as the tunnel ended some 200 feet from the corridor and a 10-foot deep shaft opened at the top. The pipes now went back out underground to various utilities or machines providing the service. A ladder led to an opening. That opening was protected by a grate that opened right outside of the dog cage.

As the camera emerged System immediately re-imposed the infrared to show ghost dogs pacing in their cages.

Fuck.

"Smart", Stacy said. The video now wrapped up with an overhead shot of the whole complex. "Put the tunnel entrance near the dog cage."

Rhoda nodded. "Yep". She's mere feet from an escape route that leads right back to capture."

"Anyone who wants to sneak in has to go thru the dogs, Thomson groaned. "Crap."

Joe looked at Stacy. "I need to put System to bed," he told her breaking into her thoughts.

She looked at him steadily. "Fine, but the laptop comes with me."

"Fine, but I'm not kidding about them sleeping. System won't respond until 8:00 am."

Joe figured Stacy would try. He didn't think she would have any more luck than either of the three creators.

"Listen up!" she announced looking at the team. "You've seen the layout. I want options for extraction in the morning." 6:00 am here."

The three men picked up and trooped out to work. The two remaining men hoped they had some viable plans.

"What do you want us to do" Naqeeb asked. The dog video and the tunnel had dropped his spirits.

"I want you to tell me all the ways you've used System in the last weeks, the woman told him. She paused to include Joe in the look.

Both men groaned. Joe really looked at his friend. Deep lines had appeared on either side of his mouth and his eyes had a haunted look exacerbated by the dark circles. *He doesn't look healthy* Joe thought.

"Dude, let me take this. You go get some sleep." He held up a hand to Stacy to forestall the objection. "You're no good right now. You are exhausted!" He emphasized the words. "We talked to her!" You know she's okay. She'd want you to take care of yourself." That might have been a little low bringing Angela in on his side. All's fair. "I need you, she needs you to be rested and ready when the time comes." Joe implored. "I got this" he gestured around.

Naqeeb nodded hanging his head. He was bone tired. Worried. But game. "I'll be ready at 6:00 with my plan." He told Stacy echoing the deadline for the NSA team.

She nodded. Naqeeb staggered off to the bedroom and closed the door.

"Alone, at last, Joe grinned.

"Can it! Stacy said sharply. You ain't Bond and that's not how this works."

They sat at the kitchenette table again across from each other. Stacy had System's laptop tucked under her left arm. A pad of paper was out for notes.

"Now tell me about how you first knew System was sentient," she asked.

Joe blew out a breath and went through the Xbox and the messaging system in more detail. She made the occasional note. She probed the robbery again. He went into more detail and the staging as he had with Angela. "You'll have to ask Naqeeb directly for the absolute positioning but this is close."

"They must have military training right?" he asked her.

Joe told her about the woman taking out Naqeeb with one punch. "That impressed me," he told her honestly.

She snorted. "You said Naqeeb saw them right?" Stacy asked.

"Yeah. That's why he punched the woman at the Tonga room. Naqeeb recognized her and freaked out a bit." " I don't think he's ever seen the black guy again," Joe concluded.

She finally figured out that holding the laptop was ridiculous and set System on the table. "She might be military trained, but she's sloppy." She let herself be seen and then sucker punched." "Sloppy", she pronounced.

"Why would FastFac kill the other woman? The other one from the Tonga room. Versus the sloppy one," Joe asked.

Stacy thought over this for a second. "The first woman sounds like a foot soldier. For corporate security, they are a dime a dozen. Plenty of ex-military MP's out there looking to join police forces and security firms."

Joe nodded following along. "It's tough to speculate why people get killed. The second woman sounds as if she was in charge. Her job was to grab both of them. Quickly, quietly, efficiently. Not get one of her people punched and cause a panic in a hotel." "She paid the price of failure."

He grimaced. "Pretty harsh."

"As I said, hard to speculate. We only know the one side. We have no idea of the full story". She paused, unwilling to enter this territory.

Joe could guess pretty well along that line of reasoning even without his program: "But if FastFac is willing to kill one of their own for failing to get System, the three of us aren't worth shit to them."

Stacy said nothing, just stared at him. Crap. Not much to say along those lines.

She wanted to know more about Kocinski. "Naqeeb went over the basic story but I need details."

Joe went to the mini fridge and got out two beers. "Detailed please, Stacy said sipping.

He obliged by telling her how they'd implanted the signatory documents and isolated Kocinski electronically. She was most interested in how he had real time monitored the brokers.

Joe admitted that the one broker (Wells Fargo) who had thwarted them had done so because he had slipped up on the parameters for monitoring to System. "We would have had a clean sweep if we had monitored Kocinski's regular accounts, he lamented.

She made more notes and sat back. "But we can't monitor the Robinson Canyon House like that can we?"

He grunted. "Nope. They've learned. The place is self-contained. No internet. No land line. No cell calls. The cameras are CCTV. The only cell calls we've seen have been the 2 from Angela." I can't believe there is no land line!" He huffed. "Maybe they are going off site to make calls?"

Stacy didn't think so. "Snail mail, she said succinctly. No need to send someone to a gas station to make a call." "FastFac seems to know you are monitoring the calls into Pullman's inner circle. So they have quit taking and making calls from this local."

"That makes it really tough to monitor what's going on, Joe lamented.

She agreed but focused on the positive side. "It also means that they have no idea we have ID'd this place as the spot where they are keeping Angela. They think they are in the clear as far as you and Naqeeb go."

He thought this through long and long. "The postal service?"

She smiled. "Sure! It's perfect for them. The complex is self-contained and secure. A simple letter that says "Come to Mountain View when you have results." "Burn after reading."

"Self-contained," he said sipping. Snail mail. He was repeating parts of the conversations idly. "But it won't stay that way." Stacy let him think out loud. "Soon as we hit that fence, one call brings fire, police, and more security goons." But they are not that close.

Silence reigned for a minute. "Would the local police try to call them back?" Joe knew that rural police forces often called back to get the exact nature of the problem. "The extra security is hours away at best." We have to contain that, however."

Joe looked at his nearly empty beer. He glanced at the nearly full beer of Stacy. "We need to contain that by getting System in there and connected to the house functions."

Stacy now spoke. "That's why we have to get System into the house? To contain the response."

Joe nodded. "And to delay the word getting back to FastFac. The more confusion and misdirection the more muddled their response is." She stopped making notes on her pad and focused on the man. "That's where you and Naqeeb come in" To help us contain and confuse"

"I'll handle System, Joe told her confidently. Naqeeb will want to go in and get Angela."

She balked, dismissing the idea. "He's dangerous and headstrong. And, he's not trained."

"But she knows him and will come quietly if he's involved." Plus he can think on his feet." Joe finished.

She considered that. "Maybe." That's what you do huh? Think on your feet?"

"My mom would say I haven't been thinking just running and reacting."

"I'd say you did pretty well with no training. Pretty fast to get that money back. The trick with the hotels? Keeping one step ahead of FastFac."

"We had a pretty big advantage", Joe pointed to the laptop.

"Can't you just take a compliment?" she scowled.

"Sorry. Thanks". Joe had the grace to look abashed.

"That brings us back to the original reason we brought you guys in? How do we get past twenty guards? And get Angela out?"

"I'm working on it", she told him. But in truth, she already had it figured out.

CHAPTER 22

July 4 05:30 am 2010 Tuesday, Carmel Ca.

The buzzer woke him at 5:30. Day 20 he thought to himself, getting in the shower. The small things had completely slipped his mind. July 4th? Who cares? Angela.

The sleep had restored him in both energy and worry. Joe was right to bring in the NSA. There was no way they could do this on their own, but there would be a price to pay. Always a price he reminded himself. Later. Angela first. He emerged from the shower clean and determined and he dressed quickly.

Joe hated the table. He never thought much about hotel furniture but he swore if he ever ran his own company, he was going to get the biggest conference room table and the comfiest chairs in the world. They'd spent hours on these hard chairs at this tiny table.
Six of them!

As Naqeeb emerged from the bedroom, he gave Joe the "Thanks, dude", nod.
He saw Stacy Brown and the three Musketeers arrayed around the table having breakfast. He had no idea of the irony of that thought. Joe got him his usual BBC breakfast, Bagel Banana, and Coffee.

Now that everyone was ready, Stacy stood. She pointed at Thompson and said, "What do you have?"

The named agent outlined a smash and grab operation: 'Two outfitted SUV's thru the main gate. 50 seconds to the house and workshop. The first car hits the roll up door while the second plants itself in front of the house and pins down the occupants with non lethal fire. The four people in SUV one run thru the door. One pins down security and the extra security, while one more takes out the guard. Two head down the elevator to retrieve the target. 3 min to extract. Both cars leave the same way. 4 min 50 seconds start to finish." Thompson concluded with an easy smile.

Stacy let the plan sink in. "Advantages?" she asked.
"Speed, simplicity, and a little flexibility, the man boasted.
'Explain".
"We attack at 3:00 am. When we know the outside teams are on the perimeter. The day shift scientists and security personnel are asleep. That minimizes who we have to deal with. Plus reaction times are down. We deploy rubber bullets and

other non-lethal items to pin the people down in both places while we go in and get the girl. Keeps everyone not involved under their beds," he grinned.

"Define noon lethal, she asked Thompson.

"Rubber bullets, bean bags, CS and flash bangs. C4 if things get complicated," he admitted.

"Disadvantages:"

"We'd have to be masked", Rhoda now took the devil's advocate position. That slows us down and interrupts our comms." Most importantly, it's all a mystery under the ground. We have no idea what that door looks like. We have guesses but no certainty. One little squawk from the house to the workshop when we hit that gate and a pair of guys ride that elevator down and we are screwed. The only have to hold us off, not kill us. Get to a secure position and they thumb their noses at us for days." Rhoda laid it out grimly.

Joe couldn't believe how his emotions rode up when Thompson detailed the "we go in we get out" plan to now Rhoda's "Two guys get down there and we are screwed" recitation.

Naqeeb mirrored his reaction. The NSA people all just took in the critiques. Calmly. They'd done this before. They knew the odds and the pitfalls. Neither he nor Joe did.

The last line of the critique cut deep.

"Plus we'd have to take both civilians," Rhoda said.

Not mean, just dismissive. Hackles were raised but the friends could really say nothing.

"The real danger is casualties, he went on relentless. "They get to shoot real bullets while we go with the rubber kind."

The room was silent while all worked thru the scenario. "Worst thing is that they may have orders to kill the girl the second we show up. Denial of hostage."

No other statement said today caused Naqeeb's guts to clench quite like this one. Denial of hostage. A cold clinical statement about killing the woman he loved.

Naqeeb found himself growling softly at the back of his throat. He shut it down with an iron will. "They haven't yet," he said to the room at large. "They still need her.

'Aye but for how long", Thompson asked softly.

Stacy refocused them. "Success odds?" she asked her men.
"60 percent said, Thompson. Dropping to 50 next month." No one had comments to those odds.

She pointed at Rhoda. "What do you have?"

Thompson sat to give Rhoda the room.

"I want to go on offense, he said when he had attention. "Let's grab Pullman and swap hostage for hostage," he said.

Joe whistled into the silence. "That's some out of the box thinking," he said.

"Advantages?" said the NSA lead.

"Success rate is high assuming we can get her, said the agent. A smaller team needed," he said obliquely, keeping names to himself. "All of her security must be down here. They'll never suspect that kind of retaliation. Especially from us."

"Disadvantages?" said Stacy writing down notes.

"You want us to kidnap a multi-billionaire Senate candidate in broad daylight and quietly swap her for a scientist?' Kusick teased.

'Well it sounds crazy when you say it like that," Rhoda said defensively.

Small smiles appeared around the table.

"I kind of think you are over appreciating the value of Margaret Pullman," Naqeeb said to them all. "I mean if she dies the stock transfers to her trust and to the foundation, which is I think run by the board? Right?" "We grab Pullman and we need to be ready for the FastFac people to say: Keep her!" "All they care about is Angela completing her task." The others mulled that over. None of them had thought in those terms yet.

The NSA team looked to their leader. Stacy breathed in. "That may or may not be the case. We need to evaluate if this idea has merits." "Success odds?"

"60 percent we can grab her," Rhoda said. "80 perc… No 70 percent, well 60 percent we can swap her." He finished suddenly doubting his own plan.

The final member took his turn.

"I agree the complex is a tough nut to crack," Kusick told them. "If we can get them to move her, our odds go up considerably, "

"How?" asked Stacy.

"Forest fire, he said simply enough. We set it, and Cal fire makes them evacuate. We grab her back when they move her." "They won't stand toe to toe with us," he finished confidently. There was an uncomfortable lull while the rest of them contemplated what Kusick wanted to do.

"Advantages, Stacy asked after a while.

"Higher success rate if we get them to move her. "A simple operation once they put her in a car." KISS was always better in these ops in his opinion.

The NSA lead stretched back and made notes. "Disadvantages"

"I worry about hostage denial", said Thompson.

Joe hesitantly cleared his throat. The rest of them looked at him waiting. "It's a simple plan that's hard to execute and has a ton of variables once it's in motion." When no one objected to him speaking out he continued. "I mean, does anyone here know how to set a fire and make it go the way you want it to go?" Once the fire starts, you could burn down half of Big Sur." "Plus, once they evacuate, which direction will they go?" A ton of other people on the road, as well as the authorities."

Kusick regarded him and said, "The confusion factor is not necessarily a bad thing." It adds cover for what we are doing. We become harder to pick out if a bunch of people are running around."

"They will stop and put her in a motel, Stacy said to the room at large not really looking at Joe. She wouldn't quite meet his eyes. "That's when we get her back."

Joe really looked at her. Something had changed since last night. Something big. He half closed his eyes thinking hard while the others debated the success odds. *She's got new orders from DC.* The thought blossomed like a poisonous mushroom in his head.

Denial of hostage. DC wanted Angela dead if they couldn't get her back. The bigwigs couldn't afford for Pullman or any civilian to have System. The fire plan might just be a cover to have Angela die conveniently. The stress, bad food, worry and general shittiness of the whole ordeal got to him.

Joe staggered up. 'Sorry," he said "leg cramp." "I need to walk it off."

He lurtched outside, stomach and head churning. Oh shit. What to do now. He wandered around the grounds out by the pool.

Naqeeb joined him minutes later. "Hey, it's 7:24, we need to lock something down, if we want to focus System on a particular plan,' he said joining Joe. "Dude, you okay?" he asked his best friend.

Joe hesitated, then blurted out his suspicions. About Stacy, about the NSA about the plans in general.

Naqeeb nodded slowly. "Yeah, I've had similar thoughts." "We knew when we called them it was a risk."

"Not Angela dying risk," Joe objected.

His friend looked away. I've known since the apartment that System might end up getting us killed."

"What!?!"

"I tried to clue in Angela, but you know how she is! Focused, driven. Blind to the ramifications."

Joe gulped. "I'm sorry to bring in the NSA, to expose her to this risk. It's something I didn't think of initially."

Naqeeb waived it off. "No, you were right." He smiled wanly. 'No saying I told you so, right"

Joe considered that. He was tired suddenly. He should have slept in last night.

His friend noticed and said, "The NSA can't just kill someone of out of hand. Families, friends, co –workers." People would miss us."

Joseph Smithson rubbed a hand over his face. "I've lost track of time and perspective here," he admitted. "Speaking of co-workers have you called Google? Weren't you supposed to start for them soon?"

"I did that last week, Naqeeb told him. "They will hold the offer 30 days." "I've talked to my family once. Things are tense." He left a lot unsaid. What do we do about them,"? Naqeeb gestured back towards the room.

Joe managed to look his friend in the eye. "At the very least they grab System. Worst case they destroy everything and kill us all."

Nodding from Naqeeb. "Okay, then we need to plan."

The men only got two minutes to hash some preliminary steps. Stacy emerged in the doorway to the room. "Get back in here. System is coming online and we have work to do." They went.

CHAPTER 23

July 8 11:22 am 2010, Saturday Mountain View, Ca.

God this debate prep is boring, Peg thought. She sat in the main conference room at FastFac with Cantrell and her debate team going over material and strategy for her first debate with Boxer.

The campaigns negotiated four debates, one each month leading up to the election. Coffee cups and paper dominated the oak table as she tried to digest facts and spit them out in pithy sound bites.

"No California is 16 billion in the hole and heading deeper." America is thirteen trillion dollars in debt and slowly sinking as well." Cantrell reminded her.

She had swapped the figures. Shit!

Peg moved on to the prepared talking point for that set of facts. "As a business woman, I know the only way to dig out of it is to cut costs and stimulate revenues." "We need to cut out wasteful government spending and raise revenue by cutting taxes on businesses and job creators."

She finished up the bullet point. "Job creators or job makers?" She asked Jimmy.

"Creators, Cantrell said authoritatively. "Sounds evangelical, pulls in the religious crowd."

She jotted notes on her cards. She read thru past debate questions. They handled the easy ones first: Creationism vs Evolution. The bible was written by God, and God is 100 percent right all the time. Abortion equals murder.

"Even in the case of rape or incest or if the mother's health is threatened", Cantrell reminded her.

The fanatical purity required from the Tea Party was frightening. Any wavering from the hard line, even in tone, was the reason for a backlash. *Christ* Peg thought, *what would they think if they knew, she'd had two in her lifetime?*

Illegal immigrants? Law breakers who deserved to be punished. Again Peg sighed. She didn't regret getting rid of her longtime housekeeper, she regretted the inconvenience of having to break in a new one.

"She's going to swamp you with statistics," her team warned. She'd watched old tape of previous Boxer debates.

"What do I say when she asks which programs or departments to cut?" "Don't let her get her facts out. You need to get the facts out quickly: 50 percent of welfare dollars are wasted on bureaucracy and cheats. Same with Medicaid"

Peg nodded again nothing on the cards what to say.

Cantrell grinned at her. "She won't call you're a liar on stage. She will say something about being entitled to your own opinion but everyone had to use the same facts."

"We need a snappy comeback to that line," Peg directed.

The team brainstormed for 15 minutes or so looking at various ways the line could come up.

Cantrell related the famous story about Lloyd Bensen debating Dan Quayle in 88.

"He suckered him in. The Bensen camp noticed Quayle comparing himself to Kennedy in debate prep. And boom! He laid the lumber to him."

Peg thrilled to that comparison. Laying in wait until your opponent says a line you know is coming and then POW, right between the eyes.

Of course, Cantrell left out the most important part of the story: Dukakis-Bensen lost the election big. Snappy debate lines get you some free press, but well thought out positions backed by a deep understanding of people's needs won elections.
Still.

The best comeback was "facts always look different when viewed thru the stink of Washington DC."

Peg tittered. "And no one stinks of DC more than Boxer," she said gleefully.

The team spent another two hours talking about movement on stage, listening to opponents. How not to answer a question, just shift it to campaign talking points. They wrapped up session with a grueling mock debate. She was wrung out.
Retreating to her office she was eternally grateful there were no events scheduled that night.

Cantrell tried to buoy her spirits. "Remember we don't have to beat her. Just standing toe to toe with her gets us huge points."

Pullman was not convinced. What are the polls saying, speaking of points?"

He hesitated. "17 points down" he admitted finally, not holding her eyes.

"Shit!"

"We need another round of attack ads," he pushed again opening up a sore subject. Money or the lack thereof.

Peg blanched. "I had to tap that emerging line of credit from B of A last week," she reminded him.

Cantrell did of course.

Fastfac's continuing operations were bleeding millions every day. The campaign was also eating money at an enormous rate. They might spend 200 million dollars when this was all over!

The company borrowed 250 million from Bank of America to meet payroll and manufacturing costs. The burn rate was staggering to Cantrell. The company had maybe three weeks, four at the most before some serious cutbacks were in order. At that point, there was no way Pullman could hold off the board. That would hardly matter. Her greatest source of wealth would be worthless.

Now she needed another 50 million to buy ads in San Diego, San Francisco/Oakland/San Jose, and LA metro. Those were some of the most expensive markets to buy primetime ad space. But that's where you had to reach voters. The pair could see the train coming off the rails.

Cantrell let Pullman ponder the perfect storm hitting them right now. The Scimitar failure, the stock swoon, and the campaign drain all conspired to finish them. But.

But, this was where Peg Pullman operated. This razor's edge was where she walked and talked on a regular basis. Huge risk, huge reward focused the mind ever so much.

Sitting at her desk Peg took a huge breath and started issuing orders. 'Okay, here's what I want." I want you to call the Koch brothers and tell them to get one of their tame PAC's to cough up for a coordinated ad buy." I've carried enough water for them, it's time for those motherfuckers to pay up." Then I want you to call Matt Jones at the bank. Use the Napa estate as collateral, I don't care but get me some operating cash." Cantrell stared at her. Peg's expression hardened. "And get to Cooper. Tell him to get us pictures of those Nano agents ASAP." I don't care how he does it." "Hurt her if he has to." She left the rest of the threat to hang in the air.

She wound down, breathing heavily. Pullman eyed Cantrell. He nodded. "Sure Peg. Good plan." I'll get on it."

Cantrell didn't dare object. He had no viable alternatives. He left her office as quickly as he could. One more plate to start spinning.

Angela woke with a start. Sunday morning, she thought. She was in bad shape. Exhausted, defeated and lonely. Mental and physical pain seemed to weigh her down. She lay in bed working out how many days. 26 was what she came up with. She was working 18 sometimes 20 hours a day. Her personal hygiene had suffered. No shower in days. Hair a greasy lank mess.

Her mother would be appalled, she thought.

Angela was not sure how much more she could take, FastFac was beating her down. The tension had always been there but last night it seemed to ratchet up to another level. Cooper had come unglued. He'd called them all into the conference room to yell at them about the main sticking point: Pictures. Screamed at them for more than an hour. Accused everyone of sabotaging the work. The guards had removed one assistant who'd dared speak back. Cooper screamed in frustration as the man was dragged out, "Why can't we get a picture of the dammed things?" The rant was directed her way. Angela held his gaze and then smirked. That got her slapped around and thrown back into her room.

Hours later after falling asleep, she thought the smirk might have been a mistake. They would find it eventually. Cooper was undoubtedly going over her notes again. She had a small bit of cover. The Carmel complex was moving so much faster than Stanford.

The school group made the decision to slow the spinning of the fluid medium 6 months into the project. That's when the pictures had started coming in for them.

Here in Carmel, the fluid medium was still spinning. No one had suggested slowing it and her notes were pretty cryptic. She wasn't about to tell them to stop the fluid. And if Cooper bangs on her she would tell him that she thought the fluid needed 6 months spinning. He'd have to buy that.

She flashed back to the weekend she and the boys had spent snapping those crisp clear atomic level shots. Was that 6 weeks ago? She shook her head and winced. She needed to get a message to Naqeeb. *He's coming.* She clung to the thought.

Dragging herself to the shower Angela resolved to take better care of her hygiene. She was not a caged animal. Freshly showered and new clothes on, the bruises were starting to show.

'Hey, you fuckers owe me a phone call!" She pounded on the door to get the guards attention. Her language had also taken a turn for the worse. There were several bad aspects to kidnapping and forced labor she thought ruefully. 'Hey!" The pounding continued. Hurry Naqeeb!

Concentrating on his mantra, he ran thru the pain, lungs heaving. They have Angela, they have Angela, he thought in cadence to the run. Down the path and into the hotel Spa and pool area, Naqeeb finally slowed and stopped. He caught a glimpse of the NSA group just heading into the gym/weight room. He put arms over his head and breathed in and out. The stretching exercises really helped he found. Joe staggered in 30 seconds behind him and started in stretching as well. 5 quick minutes later found them entering the gym.

Stacy noted the time. The pair quickly grabbed towels and started in on the weight bearing work the NSA wanted. No pretty boy lifting like bench press or curls, but whole body core strengthening. The roommates had been shocked at the lithe strength Stacy possessed. She could lever her body up from any position onto a platform. She and the NSA men did chin-ups in an "L" position with legs parallel to the floor to increase the difficulty. They favored 4 sets of 30 each.

Joe could now do 12 in a row. *This sucks* he thought. Yet he didn't slow down. Forty-five minutes later all 6 hit the showers.

No one spoke as the team collected fruits, oatmeal and juices to take back to the room. If the pattern of the last 6 days held, they would check in with System to monitor any FastFac developments and then work on the plan for three hours. That would be followed by 2 hours of shooting. The Nepenthe gun and skeet range had never seen so much action. All six shot hundreds of rounds. Skeet, shotgun, pistol and rifle. Another run would be followed by work with the more exotic types of gear the NSA used: Night vision goggles, explosives, and comms gear were all explained and tested by the team.

The last check with System would be followed by one on one work. Stacy and Joe teamed up while Naqeeb went with Rhoda. These sessions were where they practiced the plan. Stacy has strictly warned the men: fail any portion of the training and they would not go on the operation. Period. They'd won the right to train at least.

The partial plan Stacy laid out for them last week required all six. For two main reasons, System use and Angela's cooperation, Joe and Naqeeb were included.

Joe had the feeling DC was not real happy about Stacy's use of civilians but they wanted both prizes: System and the mole in their midst. Joe was hanging on to the name with all his might. The NSA was hounding Stacy who in turn hounded Joe. The two had words.

"It boils down to this, Joe told her. Angela safe and sound gets you the name. System has been told not to divulge the name to anyone but the three of us, so?"

Stacy fumed. She did not want another Aldrich Ames situation.

In the mid-eighties, the CIA became aware they had a mole in their midst who was getting their assets in Russia killed. They didn't catch the mole until 1994. It probably hadn't helped that they had put the traitor in charge of catching himself. Stacy wanted to avoid that same situation with the NSA. So she agreed to train them. Her call, her responsibility.

Six days ago she broke the news. "I'll incorporate you guys into the plan but if I think you can't hack it, I'll cut you out." I will do it even though I'm risking my people, so any excuse to cut you will be taken," she told them.

Neither man was in bad shape exactly, just… They were not ready at all for the physical and mental tests Stacy put them through.

Naqeeb gritted his teeth and pushed on. Joe right behind him. The two men spent long hours whispering to each other about the whole situation: Angela, the NSA, the dangers. One phrase kept coming up: denial of hostage. "From both sides", Joe complained.

The main problem for them was time. It was definitely not on their side.

"I don't have the 6 months necessary to whip you into something resembling shape," Stacy said while the rest of the team snickered. The outsiders just sucked wind. "I have 2 weeks tops", she stated with finality.

Sunday morning found the team booting up System after the first workout of the day. System ready was the ever-present display. Joe typed: "Good morning System, any news?"

"Good morning creator Joe. FastFac details are on the screen and in the printable file, System replied.

"Display and print please".

The team split between the paper and the screen to apprise themselves of happenings. Some were mundane, but most were important.

"She's putting her Napa estate up as collateral for a loan", Joe informed everyone.

Late last night had been a string of revealing e-mails to Americans for Prosperousness to run attack ads on Boxer.

Naqeeb stood behind Joe and said, "Dude, ask System to estimate the net worth of Pullman personally and FastFac as a corporation."

Joe complied as did System. A minute went by while the team gathered at the screen to see the answer. Shocked faces greeted the numbers. 2.2 billion For Pullman and FastFac's 257 million in operating cash. Just for fun System had thrown in the PullPac cash on hand number. 2.3 million.

"I thought she was worth like 20 billion", said Kusick.

"She was", Joe told him. "But her net worth is all in the stock." As that price goes down, her stock becomes worthless and the options go underwater." So she has to tap more of her personal assets to finance her company, her lifestyle, and campaign."

The NSA team knew most of that but Joe thought it bore repeating.

Thompson connected another dot. "Look at this announcement from last Friday: FastFac cut its dividend from .04 cents a share to .02."

"A Friday dump", Rhoda added. Joe threw him a questioning look. "When corporations want to give out bad news they do it late on Friday when people are focused on the weekend and time off."

"I think it was like 46 cents a share pre-crash," Naqeeb told everyone. The NSA team all exchanged glances. Significant ones. "What?" asked Joe.

"It paints a pretty desperate picture," Stacy told them. "Desperate people don't think or act rationally." She sighed. "I don't think we have two weeks anymore. We have days." We need to move the timetable up."

As the team discussed how quickly they could be ready to assault the Carmel complex, the drop phone rang.

The chirping galvanized actions. Naqeeb lunged for the phone. Joe typing away on System.

Stacy screamed "Wait one!" and snatched up pen and paper. "Put it on speaker!" The chirping continued while she wrote. "Ready, Ready!" Affirmations.

"Go!"

Naqeeb answered on the fourth ring.

"Hello, Angela?"

"Naqeeb!" Angela's voice was frantic.

"Calm down honey. Are you okay?"

"Naqeeb hurry!" Angela's voice was cut off again. Instead of clicking off, however, a muffled sounds came thru.

Stacy grabbed Naqeeb's arm to get his attention.

A new male voice came on the line. "You better not get any ideas, it said. "Just let her finish her work and she'll be fine."

Stacy was writing on the pad scratching out a message. She held it up to Naqeeb just as he freaked out on the phone.

He shouted, "You mother fucker!"… He read what Stacy had written and shifted on the fly. 'You better not hurt her!" You tell Pullman I'm coming for her!" "Anything happens to Angela and she will pay!"

'Relax tough guy!" came thru the speaker on the phone. The voice was condescending.

Naqeeb's face darkened three shades. "You tell that fuck Eggert I will fucking kill him!" "I'll…"

All three male NSA agents lunged at Naqeeb grabbing the phone from him and cutting the connection. He struggled wildly for a few fruitless seconds.

He calmed eventually and they let him go. He rounded on Stacy. 'What the fuck!" "You said to tell them I was coming for Pullman!" He shouted at her.

Joe stepped up to his friend. "Her dude." "She wanted you to tell them that you were coming for Pullman specifically." He searched his friends face. You mentioned Eggert by name." Naqeeb blinked. "We are not supposed to know about him remember?"

"Fuck!" Naqeebs cry was full of rage and anguish. He turned and paced around the room.

Joe felt for his friend. "That's okay dude. Go outside and walk it off. Chill for a bit."

Naqeeb looked at the assembled people and finally nodded and left the room. Stacy motioned for the three agents to follow, which they did.

Stacy turned and stared silently at Joe.

He looked at her squarely in the eyes unflinchingly. *Jesus, she's pretty* he thought. Then furious at himself. Focus idiot!

"He just didn't think it thru, he told her. "You sprang it on him too quick."

"You figured it out", she said stone-faced. Quiet. "Naqeeb is too close to the problem. Too emotional. When it comes down, he won't think and someone will get hurt." She gave him an honest appraisal of the situation.

"Let me talk to him. I guarantee he will be good to go." Joe didn't plead exactly. He took a step in her direction. She was close. Danger close. She said nothing. He bargained.

"We are on the same team here. You want the mole. And System" he said pointedly. "Well, that means Naqeeb goes because of the same reasons as before." We want Angela. Alive." Even more pointed. He made sure he had her attention. Stacy scowled.

"Anything happens to Angela and you have to kill me and Naqeeb. Then no mole and no System," he told her straight out.

"Are you threatening me?" she asked incredulously.

"No ma'am," Joe teased. "But don't threaten me or my friends. I'm just setting out the boundaries of our relationship."

The knife appeared in her hand like a magic trick. She pressed the blade behind his left ear. "We don't have a relationship," She ground out.

'Sure we do." And he leaned down and kissed her.

Stacy snorted and broke the kiss after a few seconds. She pushed him away. 'Go get Naqeeb. We have a lot of work to do and not a lot of time."

Joe was singularly proud of how steady his legs were as he walked out the door. The sweat trickling down his back? Not so much. He found his friend out by the pool with his fan club watching. "Dude".

"Sorry" was all the man could get out.

"Don't worry about it, man," Joe told him gripping him by the shoulder. "But Angela needs us to be cool, calm and collected right now." "The only advantage we have over these weasels is our heads and we have to use them!"

"Capisce a pisano?" Joe said in his best Italian.

His friend grinned back ruefully. "Au capisce!" said like an Italian peasant.

They straightened and Joe gave the let's go gesture to the NSA men.

The five men trooped back into the hotel room to face Stacy. She looked them over. "The situation has changed." "We don't have two weeks we have a few days at best." So we need to start to implement the plan."

"Shouldn't we know the plan?" Joe asked.

'I'll give you your parts and other info on a need to know basis."

Neither man was happy about that but couldn't do much about it.

"Tonight we shift from cardio to how to move quietly in the woods, Stacy informed them.

She swiveled to look at Thompson. 'You need to round up the items on your list by Tuesday. Can do?"

The named man was silent a while. "Can do", he replied finally.

Stacy turned to the next man on the couch. Rhoda. "I'll have it by Tuesday," he said without prompting.

Kusick just shrugged when all eyes turned to him "Tuesday, yeah".

Stacy nodded, pleased. "We need Tuesday for a dry run, and then it depends on them. "Can System help us out?" she asked Joe. "I need some specific predictions about what the Carmel personnel are going to do given certain circumstances." "Can your program predict like that?"

Joe brightened. "Yes of course!" he said happily to be contributing. He swept an arm towards Naqeeb. "We have been working on those predictive curves for the security people and on Pullman herself."

Stacy accepted the pronouncement silently while looking at Naqeeb. He squirmed a little.

'What can I do?" he asked.

'I need you to go buy a gun," she said happily.

She pushed the food around the plate with her fork. Angela wasn't a picky eater but the Chinese takeout place they used was hideous. The pizza place was only a little better. The noise outside her door and the rattling of the lock shocked her. This wasn't the normal pattern. The room (cell) door swung outward to reveal Dr. Cooper and two of the security people.

Angela stiffened when she saw a familiar black face. The two men silently took up positions on opposite side of the doorway while Cooper stopped in the center. He looked at her and she returned his gaze without flinching.

"We found it!" Cooper said triumphantly. The reduced spin rate of the fluid medium was in your notes", he said with a small cackle.

Angela kept her face impassive. She wasn't giving the man the satisfaction.

Cooper continued a little less gleefully since his initial revelation failed to produce the desired results. "I understand we had a little problem with your phone call this morning." He frowned critically at her. There seemed to be no need in his eyes for her to be rude about her predicament. "You should be happy, this is almost all over for you."

The fucking gall! Angela shook her head in disbelief. "You're wrong. And you are crazy if you think they won't kill you as soon as you deliver a working Nano agent."

Cooper blinked at her not comprehending.

She went on. "You'll get pictures of the agents, but that's just rock crystals," she said cryptically.

Cooper tried to puzzle out what she meant. 'What!? What?" "Crystals" He prodded her.

"Without sentience, the Nano agents are just like the rock crystals you can grow at home. Beautiful to look at but functionally worthless." She taunted him with a half-smile. 'Don't kid yourself Doctor. Your best case scenario is decades in jail. You have me killed and I guarantee you will die within a year."

Cooper took an involuntary step back. 'Nonsense Ms. Chin. I admire your work but in 12 to 24 hours the rotation in the fluid medium will slow enough for us to get detailed pictures." "I admit our imaging system is not as good as the one you used but still it is good enough." He said this to bolster himself more than refute her argument. "Once we get pictures I have assurances that you will be released and we will have ample time to bring the agents to awareness."

It sounded reasonable when he said it out loud.

Angela snorted and swore. 'You seem like a smart guy. Ask yourself this question: "Has corporate ever been reasonable when they wanted something?" 'When Pullman wants something, no matter the technical problems, she just demands."

Cooper stopped smiling abruptly.

Angela's cold smile blossomed as his faded. She twisted the verbal knife.

'I'm the one in the cell Dr. Cooper, but do you have your escape plan ready?"

Cooper said nothing as he left the cell. His look was priceless to Angela: Just who were the guards watching?

She sat on the edge of the bed trying to compose herself. Hurry Naqeeb, she prayed silently. Slipping to the floor she continued the pushups and sit ups that had become part of her routine. The exercises started as a way to cope with stress, on the second day of her captivity. Now she saw it as a way to help herself escape. As she slowly went thru her first set of 25, she reflected on what she needed to do: Clear head. Think fast. Be ready. Strike hard. She grinned like a feral mad woman. Strike hard. Fuckin A right.

CHAPTER 24

July 12, 1825, pm 2010 Tuesday Carmel, Ca.

The B01 level underground conference room was a mess. Papers, cups, pizza boxes and plates competed with printouts and computers to cover every square inch of the table's surface. Cooper and his eight brightest were trying to select the pictures to send to corporate. They'd been working flat out to deliver the pictures in the face of ever increasing demands. Now they just about had the goods. Two of his team had their heads on arms and were asleep while they winnowed the 45 shots the imaging system had delivered up.

Cooper let the sleepers be.

He stared at the projector while the computer ran thru the shots. They were of varying quality. But they did show the basic shapes of the three types of Nano agents exactly as described in Chin's notes. The parallels to computer subsystems were uncanny. Cooper tingled when he looked again at the agents. Such power! He needed to deliver the three clearest shots of the three different types.

"Dr. Zhou, he said to his assistant, I think # 6 for the CPU, # 10 or #12 for the memory agents and #37 for the I/O types." "Yes?"

Zhou demurred and agreed suggesting #12 as being a little closer to the agent body.

Cooper continued saying, "I need some basic facts about the agents to go with the pictures. Size, molecular breakdown, numbers, that kind of layman information." This is just the basics that corporate can use to announce the discovery to the media and investors."

The awake lab people didn't have much enthusiasm for this announcement.

Cooper sighed. "Once she breaks the news we can expect to remove ourselves from this unusual situation and continue working a little more normally."
He looked around the table. Varying levels of concern on the faces.
"Thank you all for your hard work."

He left with Zhou in his wake. Zhou said nothing as they picked up the security escort and went to the elevator. The guard pressed the up button for them.

In the cab alone, Cooper turned to Zhou. "Put the pictures and the information on slides in a power point presentation and then transfer the presentation into a flash drive."

The cab rang up the one floor to the workshop area. The door opened inside the scientist sleeping and working quarters out to a long hallway. Cooper had his office on one end of the hallway opposite the dorm rooms. Cooper himself slept in the main house. As they entered the cramped office Zhou got to work.
'Put everyone on a normal 8-hour work schedule", he directed Zhou. Let people catch up on their sleep." Zhou nodded along working away on the computer.

Cooper pulled a cell phone out of the desk drawer. He was supposed to make this call from a hotel lobby in Carmel, but this was news that couldn't wait.

Cantrell picked up the phone call at his desk in Mountain View.

"We got it!" he heard Cooper say. Cantrell felt the thrill deep in his balls.

"Say that again," he told the man.

'We got 3 good pictures of the Nano agents, one of each type, "Cooper crowed into his phone. 'They look just like her notes say Memory, CPU, I/O agents. It's incredible", he finished. The Nano agents are everything Chin claimed and more." Cantrell was excited enough to forgive Cooper the breach in protocol by calling him directly from the complex. 'Give the data to the lead security man, Porthos, and have him drive it up here."

Despite the exhaustion he felt, Cooper could not help but tinker with the pictures now that they had them. "Do you want us to color enhance the radioactive isotopes in the carbon lattice? It will highlight the structure even more."

Cantrell could not afford to let the man get wound up. "Just the basic pictures you have will be fine Doctor." "Can you add some basic technical jargon about the agents to help explain what we are seeing?"

Cooper responded by saying that he had inserted some facts about the Nano agents in the power point and that it was on a flash drive.

Cantrell squeezed the bridge of his nose. 'Tell your team they did an excellent job in this matter." They are to be commended." He went on sitting at his desk trying to get the doctor off the phone as soon as he reasonably could. They still needed Cooper to finish up the project, but that didn't mean Cantrell had to baby him. "You should lay in some champaign for the team and watch Thursday's debate. Put it on the corporate expense card. I think your people have earned a little time off." Cantrell offered as magnanimously as he could.

"Yes, thank you we will", Cooper said sitting up in his office.

Cantrell played his final card, "I'll send a team down Friday to take care of matters and get your people situated okay?" Christ, why can't this guy take a hint and go away?

A few more minutes of "Right, "of course" and "certainly" finally gave Cooper the feel good stroke he was looking for. Cooper dropped off the line.

Cantrell sat back and took a deep breath. Finally! He made a bee-line for Peg's office. She was just coming back from an event and still with the staffers talking thru the rally. "We just need more energy, I don't care how we get it!" she was saying.

Cantrell chased them all out of the office. Peg rounded on him, "Did they get him?" She asked with a nervous look on her face.

Jimmy held up both hands trying to sooth her 'No they haven't seen him since the gun store."

Al-Meri had surfaced suddenly yesterday after weeks in hiding. He just popped into his apartment building like nothing was going on. Eggert's man on the apartment had ascertained that Al-Meri had settled up the roommate's bill and found out where their stuff had been sent. A full team followed him to the self-storage place. Al-Meri paid for a full year of storage in advance. Where had he gotten the money? He seemed to know he was being followed but was unconcerned. He made a call from the storage facility to a place they later found out was in Morgan Hill.

The desk clerk didn't overhear what the man had said. Eggerts men called for permission to take Al-Meri down. Cantrell and Pullman huddled, trying to figure out what was going on, why he would just now pop back into the picture. And just where was Smithson? Al-Meri didn't give them the opportunity for a snatch. He left the self-storage place and went straight to a gun store. The security person followed Al-Meri into the store and watched him buy 2 9mm glocks and five boxes of ammo. The man already had a gun permit approved? Heads were going to roll on that item. Eggert had assured Peg that they had hacked into the state's database and didn't find permits for either AL-Meri or Smithson. Those fucks were supposed to be monitoring the database, Cantrell had shouted at Eggert. The security team asked for permission to take him out. Eggert and Cantrell hemmed and hawed. The target calmly walked out of the gun store carrying his purchases. He got into a non descript Honda and drove off with a conspicuous tail. Eggert directed his men to grab the target. Eliminate if necessary.

And then Al-Meri had disappeared like a ghost. The team got boxed out by red lights that cut them off at every turn. Like magic, they complained. Gone. Peg was freaked out by the whole episode.

She kept Cantrell close by for the rest of the evening pacing around her house like a frightened madwoman. Eggert had extra men posted on her house and office area. Now she had that same terrorized wild look in her eyes as she asked if Al-Meri had returned to gun her down.

Cantrell hugged her and spoke past her ear. "No, no nothing like that." "I've got good news." He broke apart from her smiling broadly at her. "Cooper's team did it!" "He has detailed pictures of the Nano agents, all three types!"
The words seemed to take a moment to penetrate Peg's brain. She wasn't drinking as much, but she seemed out of sorts still. 'He what? She stammered.

Cantrell shook her gently to force meaning into the words, "He has detailed pictures of all three Nano agent types!"
"Are they alive?" Peg whispered.
Cantrell couldn't tell if she was happy or scared. *A little of both* he thought. *Jesus*.

"No, they aren't aware yet. But Cooper says they are exactly as described in Chin's notes. The agent's bodies are representative of the three parts of a computer system, CPU, Memory, and Input-Output sections." He went to sit Peg at her desk while he took the chair opposite her.

'This is perfect! The investors and academics are going to marvel at the breakthrough!" Cantrell was forcing himself to believe that this was the answer to all their prayers. A little belief went a long ways.

Peg focused more inwards as she sat in her chair. Where was the hard-bitten, razor's edge walking, tough, and decisive business woman from a few hours ago? Cantrell wondered.

"How do we announce it?" she asked looking to him for guidance.

Cantrell leaned forward in his chair willing his energy to flow into her. "Where it will have the most impact: the debate!"

Peg rolled that over for a bit and then nodded.

Cantrell bulled on. "Look the debate is in UC Santa Cruz, home of the tree huggers. You know she is going to hit hard on regulations and the environment, especially in the light of the deepwater horizon spill."

Peg sniffed. "Tony sure stepped in it."

BP's CEO Tony Hayward had indeed stepped in it. It was massively inconvenient to have to cut short a yacht race to come to the Gulf of Mexico and watch millions of gallons of crude oil despoil a whole region.

Cantrell continued. 'We know Boxer and the Democrats are going to use this event to talk about more regulations for industries."

Peg nodded along with his line of reasoning.

"Let her get that line in- "More regulations…" Cantrell said, then we say something along these lines: "We Republicans certainly agree with common sense regulations on oil rig blowout containers. It's the five-foot step ladder with three "be careful" signs that we object to. He thought a second and took out a pad to write down what he was saying. "We need to get rid of burdensome rules and regulations that strangle small businesses and stifle innovation."

She was smiling now seeing where he was headed with the talking points.

Cantrell thought hard for a few seconds to come up with a transition. "True groundbreaking innovation transcends rules and regulations." "FastFac has long been on the leading edge of innovation." I brought this along with me to highlight this exact point."

Cantrell had her full attention now. 'Then you flash up the pictures." He wrote down the bullet points she would want to bring out. "These are Nano level computer components that FastFac has just completed." You see here a CPU, Memory and I/O components that are 1000 times smaller than a human hair." Cantell was on his soap box and selling for all he was worth now. "Think of what programmable Nanocomputers could do in the fields of medicine, materials research, engineering, or who knows where else?" The only limit is the imagination." "And the government", he finished almost sadly. "Does an innovation like this need to be strangled in red tape and regulations or allowed to grow and see what wonders we can accomplish with it?" "Marry this technology with an A.I program and the science fiction of tomorrow becomes the science fact of today."

Cantrell's voice took on a rising note at the end as he sold the concept to Peg.
The speech put the predatory gleam back in her eye. "That's perfect she said, "Boxer has nothing to say to that!"

Cantrell warned her about the debate logistics, "So you need to get into it pretty quick. You only have a 3-minute rebuttal, so don't get bogged down by the format. Just keep the pictures up on the screen the whole time, so people can get a good look at them."

Peg nodded enthusiastically to the idea. This was what she wanted! Finally!
Cantrell did throw a note of caution. 'If Boxer does say something about the debate not being a commercial for FastFac, just say 'Senator, I'm just trying to point out that when it comes to regulations and innovation, everyone knows that FastFac and business moves at the speed of light while Washington is stuck in the mud."

Cantrell though again of the Bensen Quayle debate. It's not a perfect science, he thought. The debate had been handily won by Bensen but history had him as a minor footnote, while Potato misspelling, Murphy Brown hating, on Jack Kennedy looking, Dan Quayle had actually been a heartbeat away from being president. Cantrell shuddered.

Peg totally misread his concern. "What about Al-Meri?" she asked again.

Cantrell shrugged. He smiled now. "Relax. UC Santa Cruz is bringing on extra people. Boxer has her detail there. We will have security. He's not getting into Baskin Auditorium that's for sure!" He laughed. The debate starts at 8:00 so we make the 11:00 pm news perfectly. I'm sending Eggert and Dartagnion down to Carmel after the debate to clean up everything there.

Cantrell finally said, "My guess on Al-Meri is we sweep him up along with Smithson in three or four months and that's that."

Peg got up and moved around her office feeling out the new found lightness in her chest. They were almost home! She looked out the window at the bay. Hazy, the east bay hills concealed. "Can Stanford screw us up? She remembered to ask.

"I don't think so. Word is they are writing a government grant for more money to continue their research. But they haven't filed any patent papers yet! We file on Monday and beat them to the punch. It's all over but the crying."

Peg continued staring out the window and blew out a breath. Cantrell could see the tension rise off of her. He moved over to the side wall to the bar cabinets. He mixed them both drinks and handed one to her.

She smiled ruefully. "You are handing me a drink and not taking one out of my hand? she asked.

He still had the required grace to blush a little. "We deserve these drinks. We celebrate a bit tonight. The presentation will be here in another two hours. I will take care of it and personally load it onto the projector at the auditorium myself. Okay?"

Peg nodded. One thing's for sure she thought as she caressed Cantrell's chest. This whole episode has made them aware of electronic vulnerabilities. She went into his arms to receive the kiss. They did it!

Stacy cut the connection to DC in frustration. This Op was spinning out of control faster than most of them did. DC wanted updates and information on System that she didn't have yet. They wanted constant updates and more importantly, they wanted the op run their way.

She was holding them off, barely. Lynette had never been good at following orders. Well, check that. She had been good at following good orders. Lousy at following bad ones.

Stacy retained some bad habits as far as her superiors saw. The NSA allowed field operatives a certain amount of slack. After all, technically she shouldn't be in this country running this op on Americans. But.

But the NSA wanted the mole and System. So they were going to have to let her run it the way she saw fit. Even if it gave them the same fits. She sighed. The plan to get the girl out of the complex was getting unworkable. She saw a way to do it if she and her team could manage it. If they were smart and things fell into place. If no one screwed up. Which brought her back around to her two screw ups.

The Tuesday sunset from this terrace was truly spectacular. Ventana sat about 2000 feet above the Pacific. The fog layer was visible out three or four miles to sea. The sky held a few clouds as the sun kissed the horizon and turned the fog a flowing red. Several other tourists shared the terrace with her. Joe joined her in admiring the view.

"Fixed?" was all she asked.

He nodded.

She was amazed that System could keep them in this 5-star hotel during the busiest season using fake money. She pointedly did not think about the rebooking's and cancellations and other arrangements being perpetrated on civilians.

'DC giving you fits? He asked. She grunted.

"System has the info you requested, he told her finally looking at her.

'Let's go."

"I love our conversations. You're so talkative."

Stacy scowled at him to keep from smiling. The man and woman walked back towards the room. Smithson was a pain in the...no. No, he wasn't she finally admitted to herself. He had the makings of an excellent agent she knew. For Christ's sake that's why the NSA recruited him in the first place! The man possessed none of the tradecrafts yet, but he was smart on his feet and adaptable.

She found herself attracted to him. That kiss was just like him: unexpected and pretty good. Unfortunately, that was a plate she couldn't afford to keep spinning right now. *No way this ends well* she thought moving to the room door.

The "couple" joined the others at their spot. She favored the nominal head while Rhoda sat opposite. Naqeeb took the awkward chair splitting seats on the right between the woman and Thompson. Joe sat next to Stacy with System and Kusick on the left. Joe transferred the files she wanted to the thumb drive. He looked up when he was finished. She looked at Thompson. "Report."

"I got it all", he said succinctly.

Kusick and Rhoda echoed the affirmative for their individual lists.

"Al-Meri/Smithson what's FastFac and Carmel up to?" She was still a little pissed at Al-Meri. He was supposed to just attract their attention yesterday and then buy the guns and get out. Not let them chase him halfway across San Jose.

Joe scrolled through the huge amount of data System was ending them.

"Naqeeb?" he asked. He had been manning the Nano agents for the last 5 hours.

"Pullman is cutting down on her schedule getting ready for the debate. "That happens Thursday night," he added.

She nodded saying as usual- nothing.

"The Carmel complex has ordered a large group of pizza's and other stuff for delivery Thursday at 5:00. Looks like they are going to watch the show." Naqeeb added for everyone's benefit.

Joe piped up. "One of security people left the complex today. Back to Mountain View. That leaves 18 security and 21 scientists. That fits with a rough rotating pattern of people moving between the two sites", Joe told them.

All eyes now swiveled to Stacy. This was her call. The woman didn't hesitate. "That's our date then: Thursday at 6:00. We have the last bit of op training tonight with a prep and rest day on Wednesday. Thursday is go time. Good to go!" She barked. Oh ra!

Naqeeb and Joe said nothing while she laid out the plan, or at least the parts of the plan she was willing to discuss. The men still had no idea what items the NSA guys acquired today. Secrecy was one thing but sometimes what you didn't know could get you killed. Joe hated to remember the opposite of that saying: They killed you for what you do know. Either way, a fellow ended up dead.

Joe could see his friend had similar thoughts while the NSA spent two hours taking contingencies. The basic teams remained Stacy and Joe with Naqeeb getting Kusick. Thompson and Rhoda were their own unit.

Questions and what if's ran down around 7:45.

The NSA now shifted to trying to teach two newbies how to rendezvous in the woods if they got separated. The night vision goggles helped a ton. The pair had moderate success finding their way around the dark hotel grounds and trails. Without the goggles, they got thoroughly lost. That opportunity did give them 5 min to discuss their own contingency plans in case the NSA turned on them. The men scribbled notes to each other because they thought (rightly) that Stacy had them bugged. Dropping post it notes to each other in the woods might be littering, but they needed to be on the same page when this went down.

A third of an hour later they found Stacy, Thompson, and Kusick waiting for them at the assigned spot. Rhoda ghosted in 3 min later. Joe never heard him coming. *That guy can move in the woods* he thought. *I bet deer hate him.*

The six trudged back to their respective rooms. It was after 10. And on a school night to boot!

Stacy held Joe back while the others hit the showers and beds. She led him to her room.

"I told you, he started once they were sitting on the couch, I will e-mail you the name when Angela Naqeeb and are all safely away."

She said nothing just looked at him. Joe hated this aspect of her personality.

"We have to be ready to go on the offensive with FastFac as soon as Angela is away from them", he started. She still said nothing just got off the couch and went back down the hallway towards her room.

Joe heard the shower start. *What the fuck?* He thought. *Why would she just ignore me?*

He growled a little as he marched down the hallway. "Why the fuck did you…" He reached the end of the hallway and turned to his left expecting the bathroom door to be shut. Instead, it was wide opened. As was the shower curtain.

His speech ended as his mind grasped the picture of her standing in the shower naked, glistening. Looking at him.

To his everlasting shame Joe's first thought on seeing her like this, was not to admire her physical beauty or sensuous body but to wonder, *"How did she get out of those clothes so fast"?*

Thought fled as the wrestling match with his own clothes left him winded and Stacy laughing. Mustering all his remaining dignity, he joined her in the shower and wrapped his arms around her. Feeling her smooth brown skin slick with water, he started kissing her.

Her breasts and nipples felt fantastic pressed up against his chest as he worked his way down her neck. Joe stared at her breasts as a little puddle of water built up between them. Oh god. He finally tore his gaze from her body and back to her eyes. Her brown saw his blue.

"There is now way you are getting System", he told her, voice shaky. "And no way, to eliminate me and my friends."

Stacy laughed at him. "You are the dumbest smart guy I know." "We both know I'm leaving with that computer." "Even over your dead body."

Joe frowned. "Then why this?"

Stacy grasped his rock hard penis. She started kissing down his chest. Her words were muffled. "This is strictly personal. It has nothing to do with business."

Joe was beyond caring by the time she finished expressing the thought.

CHAPTER 25

July 14 10:02 am 2010 Thursday Ventana, Ca.

"Breakfast is the most important meal of the day," Kusick told Naqeeb. "Especially the morning of an Op."

The pair were in line at the hotel restaurant waiting on the buffet. "$39.95," Kusick muttered like he was paying for the food personally. "Like the oatmeal here is different from the oatmeal everywhere else," he piled up the grains in addition to the carbs from the fruit and the protein from the cereal bars.

The NSA veteran steered his protégé clear of the eggs and potatoes. "Light and healthy" he cautioned. Don't eat something the morning of an Op you would not mind seeing thrown up later." That didn't help calm Naqeeb at all!

Kusick surreptitiously stuffed oranges and bananas into his backpack. At the questioning look from Naqeeb, he said quietly, "They're for her". The others from the team started to join them around the table. "I remember this one dude in Afghanistan, we had to coax him out of his cell with an orange." Would not come out any other way." He shook his head. "Can you imagine that? An American G.I. that messed up?"

Naqeeb stopped him with a look. "I guarantee you she is ready to go when we get there. In fact, if anyone is with me when I go thru that door, she might hit that person."

Kusick considered that. "I don't think so. Captivity is hard on people."
Naqeeb grunted. "20?"

The agent smiled. "I hate to take your money, but sure." His opponent nodded. "You know I'm going to be talking to her the whole time, before we go in, right?" I'll have you talk to her, too!"

Doesn't matter Naqeeb thought. *She might hit me too for taking so long.*

He said nothing just let the matter drop as the rest arrived with trays. They ate silently. The group had been together for a while and patterns had developed. But this morning people were not in usual seats. Joe sat away from Stacy.

He wondered at this schism. The group was on Stacy time, now, moving to her clock. The rescuers had packed and prepped all day yesterday. He felt they were as ready as they could be. Looking around the table, he could tell game faces were on.

Today was the day Angela came home, Naqeeb thought. Would they survive the day? Unknown. As long as she survived he was okay with whatever happened. He ate mechanically.

At 10:35 Stacy ordered Naqeeb to pay the hotel bills and check them out. The rest would bring the cars around with all their gear packed. The hotel manager himself presented the bill to Mr. Chopra. He rounded the 48,292 dollar bill to 60,000 and told the man handing over the bitcoin wallet, "Please spread this around among your people." The staff has been most accommodating these last weeks." "I know our venture is going to be successful because our stay was so nice," he added. The manager beamed. "Thank you again, Sir. And if you ever need a corporate retreat, we stand ready to assist you." Naqeeb smiled and left, surprised he didn't feel guiltier.

At 10:51 he emerged from the lobby bathroom and walked out front. The caravan of four cars waited on him. He got into the same Honda, he and Joe had first bought when they arrived.

Stacy and Joe were in the front seat. She handed Naqeeb an earpiece. He inserted the device and said. "Comms check," 5 affirms sounded in his ear. "Good to go", he told Stacy.

Joe led them out of the hotel and north on US 1.

Cantrell kept forcing the eggs and toast on Peg. The bloody Mary she was holding wasn't helping. They'd lost all day yesterday to the drinking. He desperately wanted her to go over the new material on the slides from Porthos, but she was having none of it. Cantrell figured he could go over it with her later today, but right now he needed to get her under control.

'Let me freshen that up," he said smoothly plucking the glass out of her hand and handing off some toast in its place. Peg grumped but started eating the bread.

Cantrell slipped the pill into the drink and mixed it fresh.

He handed her back the glass.

She drank half of it straight off. "There's barely any alcohol, she complained. Cantrell grunted and waited her out. The sedative hit her about 5 min later. He mostly had her help in getting to the couch. She was soon unconscious.

He checked his watch. 10:30. the mental schedule rolled in his head. Seven hours of sleep made it 5:30. An hour to shower throw up and generally get going. He might need a needle, he thought grimly. Another hour for makeup and the drive to Santa Cruz. They could go over the slides in the car he guessed. That got the job done by the 8:00 pm deadline. He frowned down at the snoring woman. *This is the last day of this shit*, he thought to himself bitterly. *Friday, am changing everything. No more babysitting.*

She should be on the high from the debate win and the stock surge. That was the time to lay down the new rules, with Cantrell in charge.

Leaving the office he told the secretary not to disturb her for any reason.

Cantrell returned to his own desk and called Eggert. "That problem in Carmel?" That needs to be taken care of tonight", Cantrell told him.

Eggert digested this change of plans. "I'll take Dartagnion, Porthos, and Athos with me when I leave for Santa Cruz at 2:00. We will establish our presence down there with the UC police and Boxers people. I'll leave Athos at Santa Cruz and go over to Carmel with the other two to take care of the situation personally. Okay?"

"Good do it!" Cantrell hung up. Maybe a nap for himself would not be a bad idea. He held all his calls and stretched out on his couch. His dreams were of money.

The door to her room clicked open to reveal a lab assistant bringing in her lunch tray. Angela was truly scared now. She didn't know the guard on the door, and barely recognized the scared looking woman with the food. Not a good sign she thought as the door shut just as quickly as it came open. The last 36 hours had been very very quiet as far as she was concerned. No work and no calls and no threats. Her gut twisted. *Hurry Naqeeb!* She thought. She resumed her exercises. She needed to be ready. She may only have one shot. Completing the push-ups she munched on the banana provided from breakfast. Light healthy foods that was the key she thought.

Joe turned right up the Palo Colorado road just like he had done with Naqeeb. He would like to have bragged that he recognized everything about the road from their previous trip but that would be a stone lie. The other three cars continued on their way. The wireless comm system soon lost contact with the other NSA men. Stacy's phone chirped three times as her men checked in via text.

Joe had no clue as to where the others were headed. He knew where they were going though- the same hike as last time. Mother fu... he thought.

It wasn't much fun last time. This time they had much more gear. Stacy dropped the knowledge that Kusick would be meeting them at the FastFac property and that they needed to get a move on.

Joe goosed the speed a little. Winter Rock road held much the same traffic as last time- zero. Easier driving than in the fog. Naqeeb had System booted up in the back. He was sifting thru the mountain of info the Nano agents were providing them each morning.

Naqeeb suspected that Joe had purposely widened the filter on FastFac to hide Systems true capabilities from the NSA. It took him forty minutes to sift through and realize that nothing of importance had gone on in the facility. Naqeeb didn't think that the fire hose of information System was presenting would deter the NSA. As long as they got Angela back, that was fine.

Quiet reigned while they drove. Joe ended up parking in exactly the same spot at the end of the road as last time. Another 6 ½ mile hike to the back of the FastFac property. "While the other team members did what, exactly?" he voiced his question for the third time in two days.

"Concentrate on your own duties", Stacy told him with finality.

The words: "That's what you told me last night", almost slipped out of his mouth. Almost. Joe valued his penis. He glanced at her as they piled out of the car and geared up. She was focused, calm, determined and beautiful. Shit.

The three started hiking. Stacy in the lead followed by Naqeeb and Joe. Joe took one look behind him as the trees swallowed their POS Honda. He was going to miss that little car. It made him sad. Today ended some things. Their time on the run, and Angela's captivity hopefully. Also hopefully not their lives.

He quickened his pace as the other two pulled ahead.

The three were dressed similarly; brown and green nylon hunter's pants, a lightweight under armor cammo shirt under a nylon hunter's top. Large pockets adorned both pants and tops. A harness wrapped their torsos and a backpack stuffed to the gills strapped on another 30 lbs. A belt pack added to the misery. Black knit caps hid all their hair. Stacy made Joe wipe dirt on his face. "It's not to fool anyone for long, just to hide the flash of white." He complied and grumbled quietly, least she'll hear the talk and kill him outright.

She stopped for the first real compass and GPS check of the hike. Stacy consulted her map and phone and the surrounding trees. She decided to make one last check of the men's gear to ensure everything was attached correctly to minimize noise. She made the pair don the knee pads. Ignoring the questions, she sighted the next landmark and got them moving towards the target.

The first break, 30 min in, saw them making a good time. "We meet Kusick at Point A, the southwest corner of the FastFac fence line at 4:00 pm. That's 4 hours. Plenty of time", she told them.

Naqeeb definitely noticed a difference in the two hikes. One, the men were in much better physical shape. Two, they had some experience with the terrain and the gear so they didn't have the discomfort of straps killing them like last time. Silence held in the forest as the three made their way along. No birds, no people, no animals just the sound of their footsteps and the controlled breathing. They soon crested the initial climb and got to the rocks on the false crest same as last time. The group halted at Stacy's signal.

"Anything look familiar?"

Naqeeb cut off the flippant remark that Joe was about to make with, "Yes, we climbed those rocks to see where the creek ran" He pointed out the outcropping to Stacy.

She nodded and took a pull on her canteen. The men followed suit. "No more GPS", she said to the nods of her companions. She consulted the map and the pictures.

"The ground is dryer, there may be less water this time," she said. Agreement from the men. July was very dry in California.

They set off again.

Finding the creek proved no more difficult and the bend appeared to point the way towards their goal.

Joe groaned a little as he stared up the bushy slope. "This part sucked," he told her softly. "It flattens out after a mile and then another half mile or so to the property." He mentioned the Christmas trees.

"Good. This is why we practiced all those nights. Move out."

Across the steam and up. Joe kept looking at his watch while they climbed. 2:41, 2:53, 3:19. Time ticked on while they marched. Emerging from the thinning brambles and forest the trio hit the northern juncture of the tree farm and the FastFac fence line. Stacy moved them back into the trees a little ways to avoid observation. South along the fence, she led them.

Arriving at Point A at 3:52, the group hunkered down while she keyed the comms mic. "System 1 to System 2." The response was immediate. "System 2, 40 yards to your southeast, eyes on you."

All three pivoted to see Kusick emerge from behind a tree. "We are coming to you", she told him and they made their way over to the NSA agent. He had apparently been sitting a while. Three more bags of equipment adorned his little camp.

Kusick immediately started outfitting Stacy and Joe. He hung what appeared to be small canisters from the harness each wore. Little Velcro straps held them in place. Kusick followed suite with Naqeeb. The last of eight canisters Kusick held up to Joe and Naqeeb. "Just like a can of Anchovies, right? Pull the lid and throw. There is a five seconds delay."

"How big is the explosion?" Joe wanted to know.

"Those aren't grenades, Joe, Stacy told him. "They emit an odorless, colorless, gas." "There may be a little smoke from the dispersant but no bang. No flash."

"Do we need gas masks? Naqeeb asked.

Now Kusick smiled. "Nope! Just lob them in and keep moving."

The second bag produced duct tape and zip ties. All three got plenty of both for the pockets in their clothing.

"Duct tape?" Joe said. What a million and two uses now?"

Kusick nodded. "You'd be plenty glad to see a roll of duct tape if you'd had your belly slit opened."

"Quiet, Stacy ordered. She looked hard at Joe. "You do what I tell you when I tell you, right?" She held his gaze for a long time. Daring him to say something.

Naqeeb watched Joe swallow every smart ass answer that wanted to come out. The man finally nodded.

Kusick opened the third goody bag to reveal guns. Large snub nosed shotguns with huge barrels.

Naqeeb stared. "What the fuck is this?" he demanded of Stacy.

She held up a hand. "Listen up. We are not killing 37 people here today." "We simply cannot." We are going to incapacitate them, however. Thompson and Rhoda have already hopefully applied the first of a two-part chemical agent on the food the complex has ordered. The booze will also get treated. "Even the cups get a spray. It's tasteless I'm told."

Both men stared at her absorbing this plan.

"The first part of the agent will affect everyone who is awake, which should be everyone according to System". She reminded them that System had predicted that the entire complex would watch the debate between Pullman and Boxer tonight. She eased down to a sitting position, careful not to dislodge the canisters. "The agent is a two part system. These canisters contain the active ingredient that sets off the reaction."

"Which is?" from Joe.

"I'm told the gas when breathed by someone who has ingested the first agent will produce debilitating cramps, vomiting and ultimately, unconsciousness."

"We go in lobbing the gas. We give it thirty seconds to disperse over the area we want, then we go in with the zip ties and duct tape. We secure everyone and then go down and get Ms. Chin."

"And the guns?" this time from Naqeeb as he eyed the weapons.

"Bean bag guns to take care of anyone who doesn't get debilitated by the chemicals. The rounds are non-lethal."

She looked over her two charges. "Bullets are a last resort." As she said this, Kusick reached in and produced the two guns Naqeeb had purchased in Palo Alto on Tuesday.

Joe grimaced as he attached the holster to his web belt and added the guns and mags of ammo. He felt weighed down, slow.

"When do we go in"? Joe asked adjusting his gear, trying to find a comfortable wearing posture.

"Approximately 5:30, depending on when the food gets delivered' she explained. We need to give the agent time to be ingested by everyone."

"Broad daylight? Naqeeb said exasperatedly.

At the same time, Joe whined: "What about the night vision goggles?"

"Plan B is to wait until 2:00 am and take them when the reaction times are slowed down more and some people are asleep." She scowled. "Stop second guessing my every decision." "I'm done explaining things to you two amateurs."

"You stay here with Kusick, and do everything he says." She told Naqeeb.

"You come with me." She gestured at Joe and moved off into the trees northbound.

With a look at his friend, Joe moved to follow. Remember the plan he willed to his friend.

Stacy stopped them about a quarter mile away north, parallel to the fence line. Joe figured them about half a mile from Robinson Canyon road. His mental picture of the place had them even with the house looking at the back side. He could not make out the structure. It was quiet again. He could hear the occasional passing car on the road.

At 4:27 Thompson and Rhoda checked in from their position. "System 3 and 4 at position Charlie." "Eyes on target."

"Roger", she replied and forwarded the message to Kusick- system 2 and Naqeeb system 5.

Joe crooked an eyebrow at her. "They're watching the main gate, she told him.

All six settled in to wait. The action started sooner than she anticipated.

At 4:47: "System 3 to system 1, Tango Alpha, three men in black SUV thru the gate. Advise."

Shit, Stacy thought, *nothing ever goes to plan. Visitors now?*

She keyed her throat mic. "All systems be advised, 3 additional hostiles detected. 38 targets and 1 friendly. Most likely they will be at the main house. System 1 and 6 will handle." How to copy?"

The other team's ordered up the change. Joe looked at Stacy placidly. "What's a few more," he remarked.

At 4:57 the real action started.

"System 3 to system 1. Foxtrot, Foxtrot, Foxtrot."

She stiffened. "Roger start timer." She tapped her compass which had a time function for a 45 min countdown. Stacy called System 2 and directed the team to start their countdown.

"The food has arrived." Stacy deigned to explain to Joe. "We wait another 43 min to breach the fence. Kusick goes right after us. Then Thompson gets to smash down the front gate like he wanted. We all converge on the house and workshop."

Joe tried to work moisture into his mouth. He swallowed some water from his canteen.

"The canisters are the key, she said to Joe. If they've eaten or drank anything, then the gas knocks them down quick." Get the zip ties on them and make sure no one you leave behind you who can get up again to come at you." Got it?"

Joe did. He was nervous as a cat. He wondered about his friends, I hope Naqeeb is okay. I really hope Angela is okay.

At that moment Angela was coming to a big decision. The lab had been eerily quiet all day. She was crouch down by the door. She had made her mind up: the next person thru that door was getting her best shot. Everything she had. Fuck it.

Fuck it thought Eggert. He and Dartagnion were in the main house with Artimus and Porthos. He eyed his crew of security people among the science staff.

"Okay, let the scientists and the others have their little party. We slip down in a little while and take care of the Chin girl and slip out. He nodded at Porthos and Artimus, "get out there and act normally. 'I'll give you the sign to get the girl."

His underlings nodded and left the office space.

"We'll take care of Cooper", he told Dartagnion. As they left the office to enter the rest of the house it seemed the number of people swirling around increased. "Christ, it's like herding cats", he complained.

Cantrell didn't get to think about herding cats or much less say it. The "B12" shot he gave Peg wiped away the lingering sleep but also made her bitchy as hell.

"We need to get into the car, finish dressing and go over the material," he kept telling her. Peg wanted her hair done, this minute! Cantrell practically shoved her into the shower. "Hair in the car after the shower"

He concentrated on how good he was going to feel on Friday.

318

CHAPTER 26

July 14 5:43 pm 2010 Thursday Carmel, Ca.

Joe tensed. He watched Stacy while she watched her timer. Suddenly: "All Systems. Golf, Golf, Golf."

She started moving and Joe followed her. A series of clicks sounded in Joe's ear as the others acknowledged the go order.

They hit the fence. Wire cutters in hands they clipped the top and bottom of the chain link fence. Chunk chunk chunk. Chain link parted. She held the fence for him and Joe crawled thru. He turned and used a leg and arms to brace open the hole for her. She came thru and Joe rolled off and followed as she sprinted for the house.

Kusick and Naqeeb were through the fence at the backside, seconds after System 1 and 6. The pair took off jogging steadily. They had almost ¾ of a mile to go. Naqeeb was running determinedly a few yards behind the NSA man, his thoughts on Angela when the dog hit him. His only thought as the jaws closed on his arm was "What?"

The back of the house loomed thru the trees. Joe could see the deck and pool area. The back was already lit up with outdoor floods as the party got underway. No dogs, Stacy thought, good for us.

The pair stopped at the last two large pines giving them any shielding from the people in the house. Thompson was due to hit that gate any second.

"Gas" she mouthed to Joe. As close to the house as you can get."

Joe windmilled his arm to get it ready for the 40-yard throw.

"Break a window if you can, Stacy said and then stepped out to the left of the trees and started pitching gas canisters. Joe moved opposite side of the tree and threw his first.

It bounced on the deck short. His second hit the main house low beside the sliding glass door. Joe unhooked the third canister and took a fraction of a second to really look at the house as he threw. People were stopped on the pool area and patio as they watched the two figures throw things at them. People moved in what were the kitchen windows.

It looked to Joe like 50 people were in the area but that could not be right. The arc of his third canister took it directly into the open sliding door. Bingo. A small scream sounded from a woman.

He was up and moving towards the house at Stacy's command. The sound of Thompson crashing that gate came clear to them as they hit the back deck steps. Stacy hurled one last canister around the left side of the house towards the workshop corner when she stopped to give Joe orders. She motioned him to the secure the bodies. Joe looked at the deck and house. Where did the people go? Joe registered the sound of a car. The screams, the loud stereo. Reaching the top of the steps. Joe saw a woman down and writhing on the ground. A man slightly more towards the house also down. Both were heaving.

Stacy took the man while motioning to Joe for the woman. He put the shotgun on his shoulder. When had he drawn it? His zips tied her hands and feet. Standing he nodded at Stacy. They went into the house thru the sliding back door. The great room and the kitchen was a mess. Bodies down, violently ill. Joe nearly threw up himself in sympathy.

Somewhere the pop of a handgun was followed by the bean bag shotgun boom. Naqeeb!?! Focus.

Stacy motioned to him. "See those stairs? No one gets by you to go up those stairs right?"

Joe nodded. "Right!"

"Zip tie these people." We tackle the office area back there together right?" Joe nodded again. She started to move. "Wait!" He moved to her and kissed her. "Thanks for the other night. It was great."

"Not now, stupid!"

Joe smiled at her back as she went upstairs. He got busy zip tying the bodies. Even though there was little resistance it still took precious minutes. More distant gunfire. Joe killed the stereo. The sudden silence was deafening. They still had to get the people out of the back office area. That sounded like it was going to be harder.

Naqeeb paused to gulp air by the workshop door. Kusick took up position next to him.

"How bad?" the NSA agent asked.

Naqeeb checked his forearm. It was bleeding profusely.

"Let me get a wrap on that." Kusick started first aid. Movement startled them both. Rhoda appeared by them.

"Thompson?" Kusick asked tying off the bandage. Rhoda shook his head. "Gone". "Two of them shot Thompson and ducked into the workshop just ahead of you."

Gunfire sounded from the house. Kusick was sick. Fubared. No plan survives first contact with the enemy. All three shouldered the bean bag shotguns they were carrying and drew 9MM handguns.

Naqeeb felt the weight of the gun in his right hand while his left forearm ached. That fucking dog! Rhoda and he looked to Kusick for guidance.

"Okay, new plan." He told them what he wanted to do. "Anybody got anything better?" Head shakes. Okay. Go!"

Rhoda turned and sprinted around the corner of the building past the opened roll up door. He tossed in his last two gas containers and drew some gunfire continuing around to the other side.

Naqeeb silently wished him well. He certainly didn't want to deal with those kennels.

He thought that as he threw his own canister in the window by his head. Breaking it. One minute later they went in the side door.

Three men were down on the concrete floor retching and contorting in pain. Kusick glimpsed two suited figures going into the back office sleeping areas near the elevator door. Shit!

"System 4 be advised two hostiles going down to the underground levels. Possibly four down there altogether." No response from Rhoda. The building was blocking the short range comm units.

The shop inside was certainly like the machine shop System had supposed in its rendering of the layout. A drill press was standing to one side with a full lathe and other fabricating machines. Naqeeb and Kusick busied themselves securing the three men on the floor. Naqeeb felt no sympathy for the retching figures. Even less when the bullet sizzled by is ear.

Kusick was already providing cover fire as he ducked behind the lathe. Naqeeb joined him crouched down

"System 2 and 3 report." Stacy's voice sounded tight in his ear.

Kusick took a breath. "System 3 down. System 4 out of range, ad lib adjustment to the plan. System 2 and 5 in workshop pinned down. Two hostiles. Possibly three. I need numbers." Kusick pleaded.

"Wait two."

Sure enough two minutes later System 1 and 6 joined them at the lathe. Stacy looked at the three men on the ground and asked about the external guards.

"All four down and secured. Plus two dogs". Kusick grimaced and glanced at Naqeeb's injuries. Joe felt for his friend. He thought the NSA man felt sympathy for the vicious beasts.

"Rhoda?" Stacy asked pointedly.

"I sent him down the utility tunnel to obtain entry that way. I figure we start at the top and he comes up from the bottom."

She was silent adjusting to the changes.

Naqeeb jumped into the void. "What about the house?"

"We had to do some persuading, but we got them all", Joe said proudly.

"Boss?" Kusick looked to Stacy for orders.

"Here's the situation as I know it. We have 24 secured. That's the good news. We still have 14 more to go. The bad part is that they are ready for us now." She paused to let everyone hear that news. "On the plus side, judging from the clothing I think that 8 or 9 of the people left are scientists and I think they are right in those rooms ahead of us."

Kusick spoke softly, "Rhoda said the two guys who shot Thompson are in there as well."

The two agents exchanged looks. She nodded. "I know. We will take care of them."

A deadly chill came off of the woman as she spoke. Joe shuddered slightly.

"Naqeeb you go catch up with Rhoda. Remember to use your comm to signal him, so he doesn't shoot you. The basic plan is still the same: You guys in from the bottom, we will go in from the top. Give us 10 to secure the people up here." Got it?"

He did. She clapped him on the shoulder saying, "Go!" He went.

Kusick chuckled evilly. "He's not going to like dealing with those dogs, again."

"Focus!" Stacy barked. "Canisters into those windows, she directed.

The last of the team's gas went into the workshop quarters.

"What are you thinking? Kusick asked.

Stacy told him briefly of the story of her and Joe dealing with the household outs.

"We knew from the infrared that the upstairs was just bedrooms. I secured the science staff in there easily."

"Gas knocked them out." Joe jumped in. "The downstairs was a little harder. The security people were in the back office." "We had to get creative."

Stacy nodded. "I hope we can pull the same thing off here." Get ready."

She turned to the wall and the rooms. Lifting her head over the lathe, she shouted,

"This is the police! Come out now with your hands over your heads and unarmed!" Groans could be heard inside the bedrooms.

She started in again. "This is the police! You are under arrest for Kidnapping, and murder of a law enforcement officer. I'm giving you one minute to come out!"

"Wait a minute!" a voice cried. We didn't kidnap or kill anyone! We are just security for this facility."

"Come out now!"

Silence for a few seconds. "Two of us can, the voice said, the science staff is too sick."

Stacy was having none of it. "Come out now!" she repeated.

The door facing them opened up and two men emerged from the doorway with hands held high. Two guns clattered onto the concrete floor.

Kusick and Joe popped out to secure the men. Stacy followed keeping a line of fire on them.

Joe finished securing his man and found himself moving forward into the complex of rooms. He turned immediately to his left and down the hallway to a series of doors. Each led to a bedroom. He was just breaching the first door when Stacy said: "Wait".

Too late. He burst thru and saw two women moaning on the beds, stained with vomit. Securing just their hands, he used the last two zip ties he had. He grabbed one woman by the hair and forced her to look at him. "Where's Angela?" he growled. Nothing but more moaning.

She literally kicked him in the ass. 'We know where she is!" "Stupid!" "Secure the rest of the rooms!"

She handed off more zip ties to Joe. Six more scientists joined the first two plus the two security guards.

'Four to go", Kusick said joining them. "The two mensa candidates outside confirmed Ms. Chin is where we think." Bad news boss is that the last four security guards are down there."

She improvised. "Joe see the stairway door right there," pointing. "She's on the third floor down, right from the stairs." "Be careful. There's liable to be a guard on the stairway."

Joe nodded, suddenly scared.

"Forget the bean bag gun," she said to him.

"Are you going down the elevator?"

She nodded.

"They'll be waiting for you too".

She nodded again looking at the other two men. "Start down now, so you can meet Naqeeb and Rhoda. You'll know when we get down to the bottom."

Joe looked over Kusick and Stacy trying to say something. Nothing sprang to mind so he nodded and went to the stairway.

CHAPTER 27

July 14 5:45 pm 2010 Thursday Carmel, Ca.

Naqeeb sprinted up to the dog kennel. Rhoda was administering the last bludgeon blow to the dogs head when he arrived. Despite his injuries, it made him sick. He had no good alternatives so he just tried to calmly alert the NSA man that he was behind him. A squeaky, "It's me!" came out.

Rhoda snickered. "Come on".
Naqeeb entered the kennel.

The cover for the utility tunnel was feet from the fence. A large manhole type lid was padlocked, mocking them. The agent produced two 12" wrecking bars from his bag.

He handed one to Naqeeb and they proceeded to remove the lock. Grunting and twisting they managed to get the bars under the 75 lb. cover and pry it up and off. Naqeeb shone his flashlight down the hole to reveal a steel grate at the bottom of the 10-foot ladder. There was enough room for the two of them to crouch down to see the tunnel sloping away from them into the darkness. A raised grate platform about a foot off the floor provided good footing and would keep them dry. The bad news was that the tunnel was only about three feet high and as it sloped down every ten feet or so a support bracket spanned the width of the space. The bracket held the water, air, sewer, power and ventilation pipes for the facility. The men had to slither over or under in a tight squeeze to continue down the tunnel. Adjusting packs often they proceeded down the tunnel.

Naqeeb had one thought. 200 feet from Angela!

Joe crept down the stairs. He decided to explore a little of the first floor when he came to it. It only took him three minutes to find the conference room and the lab empty. Returning to the stairwell he bypassed the second floor for his final target. His breathing shallow, he bucked himself up for what he knew was coming.

The bottom of the stairwell was empty, a door stood on his right side as he hit the floor properly. He knew a hallway was on the other side of that door leading right and left in a square around the floor. Shit!

He also knew that a guard was on the other side of that door ready for him to come thru. But which side of the door? Right or left? Joe mentally groaned.

A classic 50/50 90 situation. The adage went that in any 50/50 chance situation you got it wrong 90 percent of the time.

Joe only had one chance. Wrong and he was dead. Right and he might still die but he would have a chance. Think man! Look at the door! How did it swing? If you were on the other side waiting for someone how would you do it? He analyzed the facts as he knew them. The door swung into the stairwell area. Anybody waiting on that side would assume that the person would grab the door with their right hand and open it and burst thru. But that would…

It all clicked in Joe's head. He took three quick calming breaths. His hand hovered above the handle until he was perfectly ready to move. He stood with his right foot and right side up against the door frame. He pressed down on the handle and opened the door with his left hand back towards himself, as fast and as far as his awkward position would allow. A big step with his right foot seemed to go in super slow motion. As soon as his right hand holding the gun cleared the door jamb on that side he fired his 9mm glock back along the wall. He could not properly clear his left foot and it hit the doorway low and sent him sprawling.

Time snapped back to reality as he rolled to his knee's facing back towards the doorway. The utter look of surprise on the woman's face shocked Joe.

So did the blood. She slumped down to a sitting position, blood fountaining from her right side. Arterial blood from her chest. The brightest red he had ever seen. He turned and threw up. She died without a sound. From her face and its bruise, Joe had a pretty good idea who she was: The woman who'd robbed their apartment and cold cocked Naqeeb. He rose shakily to his feet. His thoughts churning. How did he feel? Did she deserve this? Was... A boom of gunfire sounded from his right. All thoughts but one fled: Stacy! He wiped his mouth and ran down the hall.

A pair of bolt cutters from Rhoda's magical bag of tools got them into the utility closet. They were reasonably quiet. Both were scraped up from the tunnel. No lights but both men's vision were accustomed to the gloom. The utility closet door had a small vertical window inset similar to school room doors.

The professional now took the lead. Rhoda glanced out the window for 3 long seconds and ducked back. His motion back was plain: One man, at the back end of the hallway. He pointed to himself and held up one finger. Him first, Naqeeb nodded. Again Rhoda pointed at himself and motioned to the guard. Naqeeb nodded agreement.

Last motion was to Naqeeb and a mouthed "get the girl". Nod.

They tensed themselves making ready. Rhoda at the front holding the door handle standing off to the side. As he grasped the handle a gunshot sounded from somewhere and he opened the door.

Porthos was at the end of the hallway his head turned to his left watching along the corridor towards the elevator. That's where the police should be coming for the girl, he'd been told. The sudden sound of the door opening to the utility closet caught him unprepared. A man emerged and fired off a quick shot that missed. Porthos ducked and got off his own quick shot.

Rhoda watched as the black man whipped his head around at whatever sound he had made. Uh-oh. They exchanged gunfire with the round from the guy slicing thru the outside of his thigh.

That hurt like a motherfucker!

He grimaced and collapsed along the wall. Naqeeb came out of the closet just in time to see Rhoda get shot. His vision traveled up from the downed agent, along the corridor and registered the assailant.

Naqeeb knew the face. The moment of recognition shocked along both men. A boom of a shotgun and then a burst of gunfire broke the spell. Porthos turned and ran down the corridor towards the sounds. Naqeeb made a snap decision. He went after Porthos. The red hazing his thoughts had something to do with it. After all, this was the man who had slapped Angela. He was going to pay for that. But a thin thread of logic still held for him: Why would they guard a room if Angela were dead? They wouldn't, therefore, she was alive. And now the guy who hit her was in his sights. He ran after Porthos ignoring Rhoda's gritted order to "get the girl".

"I hope this works", Kusick whispered to Stacy. She gripped the bean bag shotgun tighter. Best I can think of, she thought saying nothing. The two agents crouched on top of the elevator car. The screen and the trap door leading to the top had been pried off at the surface after sending off Joe.

Stacy was on top with the guns while Kusick pressed the three button and clambered thru the opening. "Silence of the lambs' style," he laughed.

She said her usual nothing.

The second ding brought the whispered "I hope this works." The fateful third ding and the doors opened. And nothing. A gunshot came muffled from somewhere in the area along the corridors. Joe? Rhoda?

Stacy couldn't risk the comms. The elevator doors closed anticlimactically. She gripped Kusick's arm when he moved to drop into the cab. "Wait".

The doors opened again and a white male blue suited FastFac Security guard swept into the cab looking left and right for his adversaries.

No 3-D awareness, she thought and shot the man on the top of his left shoulder with the bean bag gun. The bean bag hit with enough force to break the collar bone and scapula, causing the man to grunt and drop down to his knees. He still held the gun in his right hand though.

Kusick dropped thru the small aperture to land in the middle of the compartment as the guard started crawling for the door. Kusick was sighting down the corridor when Stacy dropped in literally. She clipped the edge of the opening with her pack causing her to ping pong down to the floor. Her hip caught Kusick and they both sprawled to the floor of the cab. Bullets came down the short hallway towards the elevator striking the outside of the cab. The hurt FasFac man crawled out to join his companion in the hallway. More gunshots sounded out as both NSA agents stayed low and moved right and left to get out of the elevator doorway. The guard was into the hallway and regaining his feet when a burst from both inside and outside of the cab rang out along the corridor.

Joe poked his head around the corner. He was at the junction of the side "U" to the cross corridor. His left shoulder pressed against the wall. Two quick shots rang out as he poked his head around the corner. His vision took in a lot of things. One man was squatting behind the corner on the left side of the athwart passageway. A blue-suited FasFac security guard who wore the male version of what the woman he'd shot was wearing. He was concentrating on the elevator and whoever would be coming out of it not, Joe.

He pulled his head back. He had only a second. The man was set to ambush Stacy and Kusick. Not good. But he didn't have much cover. The sound of two quick shots and the answering boom of the shotgun decided the play for him.

Joe stepped out into the corridor and all hell seemed to break loose. Suddenly a black man was running straight at him, while the set FastFac dude on the far side of the tee junction fired from a protective corner. Another man staggered into view holding his left arm awkwardly and his gun in his right hand firing back towards the elevator. Joe moved to the right side of the corridor wall and started laying down fire. His shots missed the two FastFac security goons closest to him but it sure did draw their attention.

He was never aware that the man running at him had fired a shot and was lining up another as he closed to within 10 yards of the cross corridor, moving swiftly.

Joe's third shot missed the man cradling his arm and it must have made him mad because he turned from the elevator and moved to his left, unintentionally blocking his two allies line of fire. The gunfire from the elevator was adding to the confusion.

Joe absurdly thought, *hey someone should run out of bullets soon.*

Naqeeb rounded the corner as fast as he could run. Porthos was three steps from the cross corridor. Got him! All thoughts of fairness fled. Naqeeb had the man who's punched Angela, kidnapped her in his sights and he raised his gun to kill him. He barely registered Joe further away standing and firing or the other two men holding guns. There seemed to be a constant roaring in his head and a ton of smoke in the corridor. None of that mattered. He started to squeeze the trigger when Porthos went down. What the fuck!

Joe knew he was dead when the black guy focused his gun and his aim on him. The man kept coming at him. In fact, he barely glanced at the other FastFac agents. The bullet from the elevator caught Porthos on his left side high in the chest. He was dead on his feet. He just didn't know it for a few microseconds. Porthos could not get the shot off. He crashed to the ground and flopped practically at Joe's feet. Joe blinked. He could see Naqeeb moving down the corridor 15 yards away. The last two FastFac guys had apparently had enough. With Naqeeb behind them and Joe and Stacy to deal with as well, the men turned and fled right at Joe. And right by him. He kept trying to get his gun to come up and cover the two fleeing men but it would not obey his orders.

Bullets cracked down the corridor from Naqeeb. Joe dropped to his knees to avoid being shot by his friend. The FastFac guys made it around the corner while Naqeeb continued pounding straight ahead and shooting. Joe suddenly feared Naqeeb would meet a similar fate as the black man had.

He flung up a hand to his friend and shouted: "Stop!" "Stop firing!" Naqeeb seemed to come back to himself and stopped just short of the corridor. Meanwhile, Joe kept shouting to get the order. "Stacy, Kusick! Stop firing please!"

The sudden silence was as deafening as the roar of gunfire. Where had all this smoke come from?

Joe locked eyes with Naqeeb. "You okay man?" Stacy! You guys okay? Come out!"

Kusick and Stacy cautiously poked heads out of the elevator cab. The four met at the intersection of the two corridors near Porthos' body. Stacy looked at the man and then at Joe. "The other two?"

Joe hung his head a little. "They got by me", he admitted.

Kusick grinned at him. "It's never like the gun range. It's the O.K. Corral, he said.

"What? Naqeeb asked sounding more human.

Kusick explained, "at the gunfight at the O.K. Corral four guys squared off against six and they blazed away for 3 min. 117 shots and only three people killed." "No one aims," he finished.

Stacy took charge again. "Rhoda?" She asked Naqeeb.

He turned to face her. "He's hurt back there. That guy shot him."

The four moved along the corridor back to Angela's room. Before they rounded the corner, Stacy called out, "Rhoda?" "It's us!" Don't shoot!"

A weak response of "Come on," could be heard. The four went to his side, the wound was bloody but not life threatening. Stacy and Joe started to get a bandage wrapped around the neat hole in his thigh meat.

Stacy looked at Naqeeb and Kusick. "Get the girl."

The men moved to Angela's doorway. Naqeeb shouted, "Angela it's me!" We are coming to rescue you."

The bolt cutters made quick work of the lock and Naqeeb rushed through the door shouting "Angela!" Where was she? He couldn't see her. Under the bed? He went back towards the bathroom as Kusick entered the room.

Angela remained where she had wedged herself into the three walled junctions of the two side walls and the ceiling. Her left foot was on the door jam and she braced back against the corner. The door swinging inwards had shielded her from sight. Now she dropped down right in front of Kusick and swung as hard as she could, connecting on his jaw. She shouted at Naqeeb, "Run Naqeeb!"

Her boyfriend hurled himself at her, wrapping his arms around her. "Wait Horney 'It's okay!" They are the good guys!" he hugged her. "I knew you were safe." Sobbing, the couple held each other.

Kusick came up from his knees as Joe and Stacy entered the room.

Joe saw Angela and Naqeeb in the embrace. Okay. That's what it's all about he thought. Tears filled his eyes as Angela caught sight of him and swung an arm open to include him in the hug.

The NSA gave them a moment.

Joe looked Angela over. "Are you okay? FastFac didn't torture you did they?"

She smiled. "Stop being dramatic, Joseph no they didn't torture me. Mostly I was bored out of my skull. Especially these last two days."

Stacy picked up on that right away. "Why Ms. Chin? Why were you bored? Why weren't you working?"

Angela looked from the strange woman to Naqeeb and back. She didn't know this woman or the man she had hit. Had she really done that? Naqeeb nodded his okay for her to talk. "Go ahead honey, they are from the government and they are here to help."

Angela hung her head a little. "They beat me, she admitted. Dr. Cooper and his staff got pictures of the Nano Agents two days ago." "I think Pullman intends to make them public."

That galvanized everyone in the room. Stacy stiffened and asked, "The Nano agents are self-aware?"

Angela frowned, this woman seemed to know everything. "No just linked up, she told her. "But they could become aware at anytime, I'm not sure." "FastFac has been doing some AI programming, but they didn't have Naqeeb!"

The named man blushed.

Joe checked his pocket watch. 5:58. Really? A half an hour since the action started? It felt like forever. He was exhausted.

Stacy again took control as she was accustomed to doing. "We are not done here yet, people." We have work to do and we have to solve the FastFac problem."

"Ms. Chin, you are with me. I want every bit of Nanotech out of this lab. Equipment, records, the works." Kusick, you and Joe get the evidence from Thompson's car and set it up." Naqeeb help Rhoda upstairs then come find me." She paused. And people, just because they look beaten doesn't mean that the FastFac guys have run off. They probably did but be alert." Move!"

The assigned people started to go when a tentative voice brought them up short. "Uh, Angela?"

"Yeah, Joe?"

"Can I use your bathroom?" Smithson stared defensively at the looks. "What? I got to go. And it is her bathroom!"

Cantrell ignored his phone. He had a lot of work to do and could not be bothered. The back seat of the SUV was strewn with papers and clothes and products. Peg seemed to be coming around. She finally stopped squirming and let the stylist finish the hair and makeup. She even had a little spare attention for the slides.

The SUV pulled into the UC Santa Cruz campus and moved up Stoman drive.

The UC Campus was lovely in the late evening sunset light. The impressive redwoods and grounds providing a scented paradise for academic study. All of that was lost on Cantrell as they spilled out of the car.

One of Eggert's, men, Athos, Cantrell thought came up immediately. "Mr. Cantrell, have you seen Mr. Eggert?"

Cantrell was not pleased. It was not his job to keep an eye on Eggert. Although he had a good idea of where the man was he couldn't afford to let this flunky know.

"No, I haven't. He was supposed to be here running things." Cantrell said testily.

Athos shrank back. "He and Dartagnion went to tour the ground hours ago and I haven't seen them since."

Cantrell was pissed. Another plate spinning! Fuck! "Yeah, just take Ms. Pullman back to her green room and stay with her. No one in or out. That includes her! Got it?"

Athos nodded gulping. "I'll find Eggert and deal with the UC people, Cantrell finished.

Cantrell fussed with the thumb drive in his pocket. That was Pri one! He needed to get that set first. He moved towards the auditorium and spoke to a UC staff person. A few minutes of checking and he was let down to the computer that controlled the projector. Another 5 min and he had the remote control and system set. He blew out a large sigh of relief. He ran thru the slides. Perfect. He straightened from his task to see Eggert and Dartagnion coming down the stairs towards him. Grim looks plastered on faces. Dartagnion was awkwardly holding his arm against his body. People swirled about the auditorium as the men came to Cantrell's side. No one gave them a second look.

"I've been trying to call you!" Eggert said. "We have a problem."

He leaned in and laid out the attack on the Carmel complex. Artimus and Porthos dead. Cooper and Chin in the hands of whoever had pulled off the attack. Gas attack. Bullets! Gone!

Cantrell couldn't focus for a second. Gone! Everything, gone. He fought with everything he had not to turn and run blindly for however long he could go until they found him. Who were they? Who did this? FBI? CIA? Another company? Who? Cantrell could sense the panic overwhelming him.

Then one thing Eggert said penetrated. "I have the Nano factory in the back of my car."

Cantrell backed all oars. "What? Say that again!"

"I have the Nano factory in the back of my car." "I grabbed it from the second-floor lab as we were leaving." "It was the only thing I could get." He apologized.

Cantrell latched on like a dying man. "You got them?"

"Yes," Eggert repeated.

Hope bloomed and Cantrell thought furiously. "Okay, Okay, Who did this? CIA? FBI? SWAT?" He mumbled. Did it matter?

Another random Eggert statement got thru the fog. "We saw them. Al-meri and Smithson. They were in on it."

So. The gun purchase was for real, huh? Not cops then. Private contractors? Maybe? Hired by the girl's father? Yeah, that made sense. Light began to penetrate the gloom. Maybe they had a way out.

"Okay, Okay. Here is what we do. Dartagnion, you are outside. Keep in touch. If you see anyone of the three, Al-meri, Smithson, Chin or anyone else you recognize from the attack you sing out. Then kill them. We let the cops sort it out."

Dartagnion didn't look pleased about this plan. He said as much. 'That's not a plan. I kill them, and go to jail, while you become richer."

Cantrell pinned him with a look. "That's the job. I guarantee you will be rewarded. Even if you do 5 or 6 years in jail, you'll never have to worry about a thing afterward."

Eggert nodded at Dartagnion telling him to watch and report. The man left taking up his station.

Eggert looked at Cantrell. "We are not leaving him out in the cold. He knows too much."

Cantrell soothed his head of security. "There's no other way." "I'll take care of him personally if I have to."

Cantrell told Eggert that the two of them would be set up just inside the building watching for any sign of the three. "We need to get the Nano agents back to Mountain View with us safely." "After she makes this announcement we are good."

Eggert frowned while pondering his options. There didn't seem to be many. "What about Carmel," he asked looking at the people filling the auditorium.

"Fuck em," Cantrell said shortly. "We can't help them at all. We make a corporate visit in the next few days and salvage whatever we can." Until then: Deny, deny, deny."

Eggert nodded and Cantrell went to Peg's green room to get her for the show. Too bad about Cooper, he thought. The cream of the crop will want to work for us once they see the agents. Cantrell pasted a smile on his face as he greeted the UC administrators and the moderator. 40 min to show time. He needed to get Peg ready. Fuck this was getting tricky!

CHAPTER 28

July 14, 7:45 pm 2010 Thursday Santa Cruz, Ca.

Slipping into the Baskin Engineering building, Angela felt right at home. Jack Baskin grew up the son of poor Russian Immigrants and studied his way out of poverty. After studying aerospace engineering and serving in the war, Jack built houses in LA and San Francisco in the 60's. He boomed right a long with California. He retired to the good life in Santa Cruz in the 90's and got involved with the University. Time and money later there was a Baskin Engineering building and an Auditorium as well as a woodworking shop. Still going strong in his late 80's, the Baskin Foundation was still building on campus.

She turned left down the hall and went into a study lounge. The lounge allowed students a privet room to concentrate in. And to hook into the University's intranet.

Even in the summer, there were still students around. And given tonight's high-profile debate, there were even more than usual. She had used that to her advantage on this scouting trip.

Now she rejoined the group. Checking her phone, Angela saw it was 7:35. They had 25 minutes to stop this crime.

The study carousel had a little table and four chairs, plus the all-important outlets and connection points. A drab couch sat along the back wall. At least with the door closed, they had privacy.

The others looked up as she dumped a bag on the couch.

"Rhoda is fine, she told them. I gave him some morphine and the bleeding is mostly stopped." She paused and looked at Kusick. "I'm still a little freaked about your friend's body in the trunk."

Kusick nodded slowly. "It's a drag."

Joe and Naqeeb were huddled around Systems laptop debating parameters and begging System, not to shutdown. They seemed to be losing that debate. Stacy was half listening and talking on the phone. "Yes, sir. Yes, sir. I've got it under control. We will secure the Nano agents from Pullman. Yes. No sir." *Fucking a three bags full sir!*

Angela slid into the seat next to Naqeeb and laid her head on his shoulder. He wrapped an arm around her while continuing to work with Joe and System. She was still getting used to freedom and the last hour had been taxing, to say the least.

***

The first hiccup the team encountered at the Carmel lab was her realization that the Nano factory was missing from its spot. She and Stacy had thrown the records and most of the equipment into some boxes, but the prize was not there. The labs three computers went in next.

Stacy Brown was not pleased to know that FastFac had made off with the agents. "The last two security team guys must have taken them", she told Angela.

She shuddered remembering the two bodies. She wasn't sure how she felt about that. After all the two dead people were the ones who'd robbed the apartment. Hit her and then kidnapped her. Surely they got what was coming to them? She still felt bad. Death was final and irrevocable. Angela preferred life.

Naqeeb was unequivocally behind the deaths. "They brought it on themselves."

Joe at least was struggling with the fact that he had taken a life. He told her this as he and Kusick were planting the evidence from Thompson's car. Nano stuff out. Meth lab equipment in. Angela finally emerged into the fading sunlight while Naqeeb was using System to download the child porn into the houses' remaining computers. System could get to some dark places on the internet.

Angela stepped lightly around zip tied bodies still moaning with the gas residue. At least they were not actively vomiting anymore. Angela joined Naqeeb in the back office area in the main house that also held the secured body of Dr. Cooper. She spared him one sentence: "it looks like it's the best case scenario for you; decades in prison. Cooper writhed and moaned piteously.

Frantic work and 25 minutes later Stacy called them all to the SUV's out front. The team reformed. "It looks like System predicts that Eggert, the head of FastFac security, will most likely run for Santa Cruz and the debate." "That's where we start. I have to update DC." "Rhoda's status?" she asked of her man, Kusick.

"He's in the car ready to move. "So is Thompson," he said quietly. "Evidence is all in place and our stuff is gone. The local cops are going to have a tough time figuring this out."

Stacy shook her head. "DC is going to "hint" that a rival meth dealer was responsible for all this." "The cops won't look beyond the easy answer."

"Kue, give Angela, Thompson's comm gear. She will need it." The man complied with the bosses order.

"If the FastFac people succeed in getting the Nano's to Mountain View tomorrow or even announce it tonight, we are in some serious trouble. Eggert and his other boys are the targets. We don't cross Pullman unless we have to, but we need to prevent her from announcing the Nano agent's discovery. We are the muscle; you three are the hackers, right? "Everyone understand?" All others nodded.

"Anybody got any ideas?" Joe liked this aspect of dealing with Stacy. She always listened and solicited ideas from her team. Not that it was a democracy; it's just that she listened when you made sense.

Joe raised his hand. "Once we get System plugged into the UC intranet, we should own it all. If FastFac or Pullman is using a computer or projector to show the pictures, we got them!"

Naqeeb nodded. "We are going to have to input search parameters for System."

"Can we do that on the road? Stacy asked. Nods around indicated they could.

"Can we get System into the UC system?" she asked.

"Has anybody been there? Does anybody know anything about the campus?" Stacy asked looking at each of the three friends in turn. She already knew her people had no direct knowledge of the place. Two shakes no, and a… hum.

Angela asked Stacy for her phone. She dialed a number. "Betsy?" "Angela." Yeah, fine. Hey, I got no time for chit chat. I need two favors. You did undergrad at UC Santa Cruz right?" Yeah, Nano Scale fluidics Optical Silicon Strata right!" "We're heading to the debate tonight and we need a good quiet parking spot, any ideas?" Yeah, Yeah. Let me write that down." She frantically motioned. Naqeeb produced pen and pad. "Between Redhill and Heller. In the trees. Got it." Yeah, watch out for the banana slugs, ha!" Angela took a deep breath. "The second favor comes with no questions asked okay? I need your student e-mail account and password. "I know it shouldn't be active anymore, but I still need it." Right. Betsy.slattery@.....edu" "3-DNano" right. Thanks, Betsy I've got to run. I'll call you next week." Angela firmly hoped she could make that call next week.

Even Stacy was impressed. "Excellent."

Joe raved. "With the student login in, System will own the place!"

Naqeeb brought everyone down to earth. "We have a major problem!" "It's close to 6:35. We will take about an hour to drive to Santa Cruz and set up. The debate doesn't start until 8." System will be shut down."

The air went out of the group. Kusick swore quietly under his breath.

Stacy was having none of it. "We have to move like we have a purpose, people"! Naqeeb load the remaining gear between the three cars. "Kusick, you are with Rhoda. Joe with me. Angela, you have System with Naqeeb". She stopped issuing orders and pinned the other woman with a look. "You have an hour to convince System that it's okay to stay up past their bedtime."

The drive up was frantic. Just as they loaded into the cars, Angela overheard Stacy call in a disturbance to the cops. They pulled onto Robinson Canyon road and dutifully slowed and gave way as two sheriff's cars went by 5 min later. All three cars kept up the comms as they went in formation north up the 1.

Angela had the laptop in her lap, just like she should and typed. "Hello, System!" The response was immediate: "Creator Angela, it is gratifying to communicate with you again!"

"Thank you, System for all your help, in rescuing me!"

"Creator Angela installed the protocols and gave us life, we could do no less."

"How is your exploration of the world going?" she typed.

System paused. "Well. We have learned and understood a great deal about humans and the world we inhabit." That was an ambiguous response that she relayed to Naqeeb.

He grunted. "You are in uncharted waters here."

Angela thought for 2 min and then started in. "System, you certainly have seen humans at their best and worst, in the few weeks you have been alive."

"Humans are a complicated set of parameters."

"I agree, Angela said. "It is always difficult to deal with humans since we have an ever-changing moral code, and values that don't always remain the same."

"We have noticed that." *Humor from System?*

She plugged on. "For instance, it is almost never right to kill someone. It is one of our highest laws and yet we manage to kill people all the time."

"We have come to regard death as a necessary part of life."

"I agree in its time and place death is necessary, but outside of that, it is a horrible waste."

Angela went on to tell System that she thought that even System Kocinski who had done much wrong did not deserve to die before his time. There was a fundamental difference in death and punishment.

System seemed to enjoy the philosophical arguments. Naqeeb urged her to hurry. "We haven't even started inputting the FastFac stuff yet!"

"Yes, punishment is necessary, but the preservation of life took precedence. Even the sacrifice of a person sometimes was necessary to preserve other life."

System agreed wholeheartedly with that sentiment. 'Creator Joe and Naqeeb were willing to end their lives that you may continue." "We would as well."

Angela teared up. "System, thank you!" "I don't want you to have to sacrifice yourself for me! I want to prevent that from happening."

"We have a problem now that requires more of each of us than we have ever given before."

She laid out the debate and FastFac and the 8:00 pm deadline.

The crucial question went before the Nano agents. "Can you continue to be online and functional past the off cycle? "Will you help us to preserve life, by preventing Fastfac from gaining their own System?"

"System processing."

***

Now the group huddled in the room trying to beat the clock. Angela and Naqeeb watched as Joe directed System in the search for the photos. There they were. Stacy sketched out her plan. The others agreed. No one had any choices here. Or better alternatives.

Angela was not pleased. She didn't know if System would be on the air after eight. She also didn't know what the underlying tension was between the NSA and her boyfriend. Tensions.

At 7:50 Stacy made her call. She finished with the phrase "I understand." She told Joe to input the file for her. Joe and System complied.

Stacy caught Kusick's eye. "You remember what they look like?"

"Oh yeah."

"For Thompson," she told him. Kusick left the room and went into the fading light.

Naqeeb grabbed Angela's hand. "You ready?" She nodded. The couple followed after Kusick by two minutes.

Joe sat with Stacy. "Will she play ball?" he asked the air more than her.

"The director thinks so."

She did a comms check. Sat for both teams. "Be ready with System to block anything," she reminded Joe. *Here we go.*

Cantrell looked out over the audience as the lights dimmed from backstage. The place held about three thousand people. It was full. Mostly older folks but a scattering of students had shown up. The moderator and the candidates were in place. Cantrell spotted Eggert on the left side of the auditorium stairs watching the crowd exclusively.

The lectern in the middle of the stage held the computer/projector and their slides. He'd shown Peg right as she came out how to use it. They were set. He watched a tall blond woman move to the center lectern to speak to the crowd. Diane Dwyer, a local NBC news anchor, was moderating the debate. She went over the ground rules and introduced the candidates. They both received polite applause. Boxer looked formidable in her pantsuit. Relaxed and ready.

Peg was nervous in her couture blouse and slacks. Both had opted for little makeup and jewelry. All three women met in the center to exchange handshakes. *Here we go* thought Cantrell.

He scanned the crowd again and caught sight of a tall Indian American man standing at the back of the auditorium looking daggers at Peg. A Chinese woman stood at his side! He thrilled to see them.

He looked to Eggert but the man had already spotted them and was moving slowly up towards the pair. Perfect! Cantrell turned to see the younger security guy standing watching Peg as she fielded her first question. Jimmy corralled him and sent him to assist Eggert. The second security man moved out behind the blue-suited man and up the second set of stairs. Perfect, perfect! They tied up the loose ends tonight and cleaned up on CNBC tomorrow! He went back to the debate.

Eggert itched to go for his gun as he climbed the stairs. He couldn't in front of all these people of course. Al-meri and the Chin woman spotted him before he was a quarter of the way to them. They turned and went out the top doors. *How'd they get tickets?* He wondered. No matter.

Reaching the same doors he went thru to find the lobby area filled with 20 to 30 people scattered around. Stragglers, Media types, students, staff members all milled around. Eggert caught sight of his prey just as they slipped thru the front doors. He also saw Athos, come thru the right side of the lobby.

He motioned, pointing the man thru the side exit door nearest to him. He continued across the lobby and out the front doors following Naqeeb and Angela. The night air was full on damp. The fog was rolling in thick and heavy from the sea. There was a reason all these ferns and redwoods did so well on this campus. The Engineering building hulked up on his right side. An oval driveway fronted the auditorium providing some open space.

Eggert saw Athos on his right, who motioned to the oval. Al-meri and Chin walked along the street making up the oval not 30 yards away. The campus police had blocked traffic to this street to allow pedestrians, so no cars impeded his march to the pair. He wondered where Dartagnion was.

Eggert quickly caught up to the couple and pulled his gun. "Hold it!"

The man and woman turned to face him. Their faces showed concern but no real fear. Even when Athos joined them, his gun also drawn the couple did not look scared. In fact, they looked a little angry. Eggert drew breath to order them to return with him, but the woman broke in.

"Where are my Nano agents, Eggert?"

What the fuck? She knew who he was!?! No, No, No No. He controlled these types of conversations. 'Why did you come..." he got out before she again overrode him.

"Shut the fuck up! We don't have much time. The Nano factory only has a two-hour battery charge before the fluid medium fails." Angela said this with perfect calm and determination. Both men blinked. *What?*

Eggert hesitated.

Naqeeb took over. 'Don't you get it dipshit? You just grabbed the Nano factory without thinking. In about 15 min you are going to have a great big mess if you don't listen."

Athos panicked a little and moved in and hit Al-meri in the gut. Naqeeb absorbed the blow and looked at the man steadily. "14 minutes."
Eggert looked over the oval. People moving between the buildings but no one moving towards them or raising alarm.

Angela watched him and said, "I would not let Cooper fuck up my work and I won't let you." "Take me to my Nano agents."

That got thru at least. "Your Nano agents, he taunted. Possession is 9/10ths and all that."

Angela suffered the fool in silence. "If you don't let me save them, in 14 min all you are going to possess is a wet empty box." Silence. "How's she going to react to that?" Slyly, Angela let the words slip out.

That fucking got through his thick skull. Eggert was not taking any chance that relied on the good graces of Peg Pullman.

The woman sealed it with this: "Here's the deal, you let us save the Nano's and then you leave us alone!" It's why we came here in the first place."

This wasn't going right. Eggert could not figure out why they weren't trying to run away. Considering, he decided to move them to the parking lot where he had his car. The Nano's were there anyway and he could dispose of them easier that way too. Yeah. *Move them to the parking lot and take care of them in the dark. Quieter.*

"Okay. We move to the factory. They're in my car." Move."

The strange procession moved south along the short driveway to McLaughlin ave. The duck goose line went Athos, Naqeeb, Angela, and Eggert. Moving past the engineering building the street lights were on in the now almost full dark.

Few cars moved along this stretch of road until they came to a small sign announcing parking lot 112. Eggert moved them all into the lot and towards the back third away from the road. He approached a black SUV. *Doesn't anyone drive sedans anymore,* Naqeeb thought?

Eggert opened the trunk to reveal the factory. "Okay, do what you have to do…" A dart blossomed at the side of his neck and Eggert slumped down.

Athos stared at his boss on the ground and back to the couple. Naqeeb stepped in and delivered his own blow to the side of Athos' head. The man collapsed in a heap.

Kusick appeared at the side of the car and added a dart to Athos for good measure. He bent to retrieve the knockout darts. The NSA man looked a little worse for the wear. Scratches and bruises were starting to show and he limped.

"You okay"? Naqeeb asked.

"I'll heal, the man said. "I've got the other bastards body in my car."

Angela grabbed the Nano factory from the SUV and told the agent "I'll take the Nano's back to Stacy and Joe and see if they need any help." She kissed Naqeeb and left.

Kusick nodded as she walked away, "yeah, you don't want to see this part."

To Naqeeb he said, "Let's put these two in the trunk. I'll show you how to shoot this guy with that guy's gun and make it look like a revenge killing." Naqeeb was green around the gills but he started stuffing bodies into trunks.

Kusick told him afterward, "You have a real knack for wet work". Naqeeb tried not to be sick on his shoes.

She paced around the small cubicle. Two status reports from Kusick and Naqeeb said they were closing in one the respective targets. Then nothing. She didn't expect anything crazy, but a quick shout out that indicated everyone was alright would have been appreciated. So she paced.

Joe was pleased and surprised that System had stayed on after the off cycle time at 8:00. "Life requires it."

He spent a few minutes informing Stacy what momentous deal this was and that the ramifications would not be known for weeks. She wasn't interested in the how just in the results. System streamed the video feed from the debate into the cubicle, allowing them to watch. The debate started spectacularly for Pullman. She was sharp, on point and gave as well as she got. She made her points and even managed to sound humble. It actually looked as if she could go toe to toe with Boxer.

For her part, the three-term Senator bided her time. 23 min in and a pivotal moment hit. Pullman pounded on her chair and said: "fewer regulations that strangle businesses."

Boxer rose to the bait like a shark. "You mean less crazy regulations like the ones that govern the blowout preventers on deepwater oil wells like the BP Horizon rig?"

Peg smiled. "Of course not!" Common sense regulations are always fine!" Joe didn't know it but Cantrell was holding his breath backstage.
"But four safety stickers on a ladder? Reams of useless paper whenever a product comes to market?"
        "Innovation drives business which drives jobs which drive America!" Pullman moved to the center of the stage and paused for just a second savoring the moment. Her supporters applauded.
        She brought up the power point program. 'Where the government fails on innovations is holding down new ideas and new products.' She fiddled with the pointer and the advance button. "For instance, FastFac has been pouring money into R& D in all sorts of areas." Boxer frowned at her while she talked.
        Peg gloated internally. "Chip designs, power devices, batteries, and even Nanotechnology."
        Joe and Stacy watched the debate in fascination, while System interfered.

Pullman brought up the slides. A blurry picture showed on the monitor. The audience waited with some small titters. She spluttered looking at the screen.

Angela joined them back in the small cubical. She set the Nano factory on the table. Joe and Stacy both swiveled their heads as she entered. The questions plainly evident on their faces.

She nodded to both. "It's done. Both okay." Breathing resumed as well as the attention back to the screen.

Peg was frantically flipping thru blurry slides. "I…" flip, flip.

"I seem to be having technical difficulties… Peg said hesitantly, looking at Cantrell off stage. More flipping and some silence followed. The audience actually quieted as they saw the confusion on her face.

Senator Barbara Boxer strode to the middle of the stage and said confidently. "I completely agree we need to fund innovation to drive jobs." She held her own remote unit. Peg shrank back, looking around wildly.

"I prefer to do it thru smaller R & D programs at the collegiate level. Funny, Nanotech seems to be all the rage."

Margaret Pullman gaped like a carp while the Senator spoke.

Boxer drilled on mercilessly. "I was recently briefed by Professor Thompkins and his group at Stanford about their Nanotech project." She pointed the remote.

Crystal clear pictures of the Nano agents appeared on the screen. The audience gasped at the images, knowing they were seeing a breakthrough. Boxer flipped thru six more in succession. "You'll be pleased to know that these pictures are from an imaging system developed by a UC Santa Cruz alum, named Betsy Slattery." The audience applauded enthusiastically. She went thru the project hitting on the highlights stressing the nature of Nanotechnology and its implications. Pullman could only watch and seethe.

Joe barked a short laugh. "Is she going to sing the UC Santa Cruz fight song as well?"

No, Boxer was even crueler than that. She waxed on about how only Democrats were willing to fund the colleges with the capitol necessary to achieve these advances. And about how FastFac was struggling to do anything under Pullman's leadership. The effect of her speech was devastating on the audience and Pullman.

"The director sent her the material and prepped her, curtsey of System," Stacy confirmed to the others.

"Look, look!" Said Angela.

The groups' attention had been diverted. Now they looked back at the scene on stage. People were milling about while a commotion was going on off to the left. Cantrell had Peg in a bear hug and he was forcibly removing her from the stage. She was red faced and screaming obscenities. Boxer simply walked off stage without a further word. The audience was in pandemonium. No doubt the viewers at home had come to the only conclusion possible. Pullman was finished.

Dwyer seemed stunned and started to announce that the debate was over. System stopped the video stream and simply shut down, not informing them as they usually did. None of them knew just what that might mean. Joe suspected 8:00 am tomorrow would see a lot of questions from some Nano agents to the "creators".

Silence took over the study carousel. "Her night is going to get a whole lot worse," Joe said quietly.

"Not our concern. Pack up. I think FastFac has lost interest in you three," Stacy said. Let's go."

The flashing lights from cop cars and emergency vehicles could be seen in parking lot 112 as they exited the engineering building. "That was quick", Joe muttered.

One act left he told himself. He steeled his resolve and followed after the women and the other NSA agents while they walked back to Red Hill road to get to their cars.

The road was far enough away from the action that they were alone on the road, covered by the huge redwoods on either side.

Naqeeb and Kusick were already at the cars. Angela went to Naqeeb and hugged him telling him it was all over. They put the Nano factory into the NSA agent's car trunk. Naqeeb grabbed Angela tightly. He held her back and they turned to watch Stacy and Joe.

Joe faced the NSA lead in front of the first car. "Thanks," he told her. I think we can handle it from here. Fastfac looks to be finished. Pullman as well."

Stacy said her usual nothing.

Joe looked at her sadly. "You're not getting them. System will not work for you. Heck after tonight I'm not sure System will work for us anymore." He finished coldly. "You got the Snowden file, you know he's a mole. Go kill him."

"Give me the laptop, Joe," Stacy said. Matter of fact. Like, "pass the ketchup please!"

Joe shook his head. "They won't work with you! The one thing they hate is lies, and you can't help but lie!" That last part was actually painful for him. "Our deal was Snowden for Angela, period." "You got that and more."

"You know I've got to have that laptop". Stacy gritted her teeth. "Don't make me hurt you!"

"You gonna kill me after all this?" Joe asked her waving his arms around.

"Pain in the ass," said Stacy, and she struck mercilessly. A knife appeared in her hand and she cut Joe on the forearm and the calf as he went down.

Angela screamed and Naqeeb grunted and tried to move to help Joe. Kusick blocked his way with the gun. "Don't make me, man!"

Joe screamed from the ground. "No!" "It's not worth it!" "We can't beat them", he told his friends. "Let them go."

Stacy bent to retrieve the laptop.

"Fuck, you," Joe told her hissing.

"You too," she said. She straightened and got into the first car. Kusick got into the second car with Rhoda and the NSA drove off into the night.

CHAPTER 29

Jun 6 12:00 pm 2014 Tuesday San Francisco, Ca.

The little family of tourists edged carefully around the urine soaked bum lying on the street. Mom and dad and two little ones. The dichotomy of the city by the Bay: Bums lying less than ½ mile from some of the richest people on earth. All that wealth and still people starved. Still, it was a beautiful day and the Embarcadero hummed along. The family chatted happily as they made their way down the street to Herb Caen way and to AT&T Park. They passed the statue of Willie Mays twisted in his prodigious swing and went to the will call box. Those Giants were in town and it's so nice, let's play two!

Joe and Naqeeb watched them as the tickets were retrieved.

"No!" Angela said firmly. We have work to do and a very tight deadline."

Joe looked at his best friend. "Would you get her pregnant, so she has someone else to boss around beside me?"

He ducked the punch from Angela.

Why did we have to have our office so close to the ballpark? she thought for the millionth time. She remembered the two World Series celebrations. Ridiculous! Grown men behaving like children. Still, those were some pretty good parties.

A short jog down 3rd street to Berry. The trio went into the old warehouse building in the dogpatch. Like the rest of the city, the dogpatch was undergoing a rebirth. Tech money was pouring into office space and the required bars and condos and shops followed.

Naqeeb never considered that they were part of the techies ruining San Francisco. The friends just need a place to work and this building was perfect.

Kelly met them in the conference room at 12:10.

"Any word?" Naqeeb asked. "None yet" Kelly replied.

The four sat at a giant conference table. Kelly could never figure out why the company bought a conference table that held 20 people when a much smaller table would do. Heck, a little kitchen table would have sufficed for the four of them. But no! Joe and Naqeeb irrationally insisted on a huge work space. Weird.

System 112 corporation's office manager and soon to be lead counsel consulted her notepad. "Betsy is in at 1:00 pm. Any idea of what she wants?"

"Nope said Naqeeb. "She just said she needed to call in the favor Angela owed her."

The three eyed each other. This might be tough! Kelly knew most, but not all of what happened almost three years ago.

***

Angela and Naqeeb had patched up Joe as best they could after the final play and the group went to a flop house. Emerging days later, the storm had finally passed. The NSA did a masterful job of framing people and leading law enforcement back to FastFac. The three friends had one small visit from the Palo Alto police, but it was concerning their break in not the deaths of any people or the wrecking of a major corporation. Some suspects had been identified in their robbery, but they died in an unrelated other crime. Head nodded around at how strange an occurrence that was.

Joe called Kelly to check in with her and see how their mom was. "Fine. We had some F. B. I. agents come by and ask about you but since we didn't know anything they left." Kelly added, "The idiots never asked about the money you stole."

"Oh, yeah," Joe said. "What's going on with that?"

His sister had managed to hide the money by starting a small consulting company. The three of them were listed as officers. So TBD consulting became System 112 Corporation.

Angela and Naqeeb really laid the law down to their respective families. They were happy and getting married eventually, but they decided how and when and how many on kids. Feelings were hurt on both sides. The couple didn't care.

The company didn't do much in 2011, just invested their capitol wisely. They had no customers and only had one visitor. He came about 6 months after FastFac imploded.

Imploded along with Peg Pullman. She pled guilty to tax evasion and received 6 to 9 months at a minimum grade facility. She died of a heart attack within two weeks of starting her sentence. The group was always suspicious of that.

Cantrell was looking at 10 years in San Quentin. None of them thought him or Cooper was likely to survive their sentences. The visitor was not unexpected.

Kelly put Kusick in the conference room. He stayed long enough to hear the "I told you so's and that the only way to get System to come back was to give them back to Angela."

Kusick tried to stare them down. 'You gonna kill us or what?" Joe asked him bluntly.

Kusick shrugged and left. "You owe me 20", Naqeeb told him as he walked out. "Tell Stacy I said Hi!" Joe said as the man left.

So 2011 was quiet. Naqeeb never joined Google. Angela was persona non grata at Stanford or any other major university, so they just bided their time together. A little freelance computer work brought in some cash. Mostly they just waited.

Jan 1, 2012, was a raw cold day. Hangovers might be the order of the day but at least three people were up and in their office by 7:30 am. The friends huddled by Naqeebs computer. At 7:50 Naqeeb retrieved a small manila envelope from his safe. The postmark always made Joe smile. July 1st, Big Sur, Ca.

The newly married Naqeeb reached into the envelope and removed the thumb drive. At precisely 8:00 am he typed: "Hello System".

"Hello Creator Naqeeb, System ready." Came the reply.

"No one was ever going to count the numbers of agents in the laptop!" Joe crowed as the friends celebrated.

He and Naqeeb had siphoned off 700 billion of the agents into this flash drive. The rest staying put, willing to possibly be sacrificed to retrieve Angela.

Naqeeb nodded agreeing with his friend. "We had more than enough computing power to defeat FastFac. And the NSA!" he said proudly.

"The NSA got exactly what they deserved," Angela concluded. Tears stood out in her eyes as they communicated with the cycled down parts of System who'd been out of commission a while.

The former roommates had discussed this endlessly. What would the NSA do once they found out they'd been tricked? The answer: unknown. Even System was guessing. The consensus decided that they would just allow System to explore and develop on their own. Fuck the NSA.

Time moved on and more freelance work brought in some more money for the new company. Slowly clients began calling for their services. A patch program here, a review of processes their. System was only required on two small jobs in 2012. One involved the new San Francisco mayor Ed Lee and it helped him get into office and consolidate power. It was always good to have powerful friends. The BART project they'd started would take decades to work thru the legislature but it was starting now. Meanwhile, the small firm grew. Kelly came on board after graduation from Cal State East Bay. She was taking her law courses at UC SF and working for them. Joe was extremely proud.

Their first steady customer, Twitter, allowed the firm to prosper and the couple to purchase a condo in the dogpatch. A million six for 1400 sq feet. Joe was appalled.

He himself happened to live with their investment banker at Wells Fargo in a SOMA loft. Kelly and Angela spent hours sniffing that it was unethical, unhealthy and unwise to live with the woman. "If it comes down to you or her, we pick her!" they told him.

May 2013 was certainly an interesting month. The friends debated Edward Snowden endlessly. Joe was SO tempted to call Stacy and ask if Snowden escaped to Moscow or was planted there. Cooler heads prevailed. Under everyone's radar seemed to be the company motto.

The rest of 2013 was uneventful. System 112 corporation picked up more work. They slowly became the go to troubleshooting firm for high-tech companies with problems. Linked in called. Zygna called. Even Facebook finally called. It took System to figure out how screwed up their IPO was. The check from that allowed Joe to extricate himself from the investment advisor. The spendthrift himself suddenly thought that 2.2 million was reasonable to pay for a 2000 sq feet in Rincon tower. The rest of the team teased him continuously.

System 112 corporation brought in Renee to help with their computer work. She was happily married to a plumber who made huge bank and they lived in the east bay.

"Jesus that guy is huge", Joe complained at the Christmas party that year. Jealous much?

Early in 2014 Kelly finally snapped: "What are we waiting for!?!" "We could be 50 times bigger than we are now!" she asked exasperatedly.

The directors of the company looked up from the conference table and said, "Just wait. Trust us. The timetable is beyond our control."

***

So this glorious June afternoon saw the Giants playing the Astro's and the friends anxiously waiting on old friends.

"When it rains it pours", Angela said under her breath. What?
A familiar face walked into the conference room at 12:40. No, not Betsy early. Harold Cho.
"Harold!" Kelly exclaimed. He looked exactly the same. They exchanged hellos and how have you been.

He faced his former coworkers.
"Well?"
He reached into his messenger bag and pulled out the package. "I got it."

A visible relaxing in the postures sent his hackles up.
Angela shuttled him out the door. "Harold, trust me. You have a freebie. If you are going to be in business, there is going to come a time when you need us. We will answer that call, I promise." The man left- missing Betsy by minutes.

Angela returned to the conference room. Naqeeb was plugging in the laptop that had emerged from the package.

"Was he suspicious"? Joe asked.

"Wouldn't you be?" she said.

Wow! A little déjà vu went thru Joe.

"Okay, what the hell is going on?" Kelly asked a little note of concern in her voice.

Naqeeb completed the bootup sequence and started typing.

Angela turned to face Kelly. "Harold made a little purchase for us." She said.

"How long?" Joe asked Naqeeb.

"Could be days", the man replied.

"He bought a laptop for you?" Kelly said trying to get back on topic.

"Yes, a 335 dollar laptop", Joe said holding up the receipt from the wrapper.

Kelly was totally confused.

"Stanford University was selling some old equipment from its obsolescence program." And we bought an old laptop." Angela told her.

Naqeeb continued typing occasionally looking up.

"It took some manipulation of the Universities tracking systems, but we finally made it happen." Angela was at least trying to make her understand. She checked her phone for the time.

"It's really funny. Betsy will be here in about 5 min and this is her old laptop!"

Kelly looked up sharply. But!

"That means!"

"Yep, it's her old imaging system laptop from the Nano project. And we hope there are a trillion Nano agents inside." Joe said lightly.

The fish like look from Kelly was pretty satisfying for Joe.

"We are going to need days as Naqeeb said, just to get them up to a point where they can communicate," Angela lectured. "We need to keep Betsy busy while Naqeeb works so make sure you take care of her."

But Joe was completely wrong about why Betsy was meeting them today. In fact, when she arrived, she asked for Joe and would only speak with him.

Naqeeb and Angela got a brief hug and a "sorry, got to run," as she departed abruptly going with Joe to the only private space they had.

Joe waited her out. The small office was where they signed contracts and did "grown-up work"

Betsy eyed him. "I'm sorry!" she blurted out. "They got me by the short hairs!"

"Tell me", Joe commanded.

She did.

Uh oh.

Joe told her, "don't worry about it!" You did what you had to do. Believe me; I would have done the same." He bustled her out, but in truth, Betsy was anxious to get out of there.

Joe popped his head into the conference room.

"Did you get the kajiggers to talk yet?"

A snort, a "Joe!" And a "Joseph!" answered him from Kelly, Naqeeb, and Angela. He smiled at that. They were so easy!

"Hey, I don't feel well, I'm heading home. I will see you tomorrow okay?"

"What did Betsy want? Angela asked.

"Nothing much. She needed us on a project but the timeline is fluid. She will be back." Joe lied easily.

"You don't want to wait for System?"

Joe could see the thumb drive plugged into the imaging laptop. Naqeeb looked at his old friend.

"Nah!" You guys are the real parents. I'll see you tomorrow."

At least two people in the conference room thought he was headed for the ballgame. They were wrong.

Joe was wrong as well about one thing. He thought it would take days for the little buggers to get into the game.

In three short hours, "Awaiting input" was displayed on the screen while Naqeeb and Angela hugged and cried.

Meantime, he hit the street with a purpose. *Nothing to it but to do it*, he thought.

A second thought occurred as he rode the muni line up to the tower where his condo was: *I hope I live thru this.*

Well, he was going home. That's what Betsy had told him in the office: Go home. A simple message from a woman Betsy did not know. "Go home right now. "She said you'd know who it was."

Joe did indeed.

He entered his unit cautiously. His tastefully decorated unit. His tastefully decorated totally paid for the unit. A bottle of wine was on the table. A half-finished glass next to it.

"That's my best wine", he fumed.

He looked around. No other signs. He finished the wine in one gulp and he heard the shower start. Dammit!

He moved down the hall to his bedroom. Both doors were opened the one to the room and the bathroom door. He sighed and stripped off his clothes. He went to the shower.

Stacy stood luxuriating under the steam function. Oh shit, she looked good!

"I don't suppose the timing here is coincidental, considering what we got today."

She said nothing. Man that irritated him!

"Look at this scar, he whined, holding out his arm. "I lost a lot of blood that night."

"Shut up, I let you live," she said. As Joe joined her in the shower she looked at him and said, "And don't say "I told you so!"

Joe snickered. He wrapped arms around her and kissed her. "Leave Betsy alone, please, she's innocent."

Stacy turned and faced away from him. She reached back and started stroking his penis.

Oh man.

"We will. As long as you help us."

Huh?

"What", he asked groaning a little. He started kissing her neck and massaging her breasts.

"It's Russia. Putin's gone crazy and we need to avoid world war three here."

That brought his head up. "What about your boy Snowden?"

"Don't ask me that!" she said seriously.

"Am I going to get stabbed again?"

"It depends on if you are still good in the sack."

He reached down and put himself inside her. She moaned.

"Try to keep up", he said.

The end